Keeping Warm

Keeping Warm

A guide for wintertime

Edited by Barbara Kelman
and Jeanette Mall
with Ellen Leventhal,
David Sachs
and Martina D'Alton

New York London Tokyo

Contents

Introduction

This book is about coping with cold weather. A group of experts in various areas have gotten together to share with you what they know about keeping warm. Within these pages you will find a wealth of general information and lots of practical advice and instructions for projects to protect you from the ravages of winter.

If you own a home, you have come to the right place for advice about insulation: Is what you already have enough? If you need more, how much? what kind? What will it cost? Can you install it yourself? If leaky windows are cooling off an otherwise adequately heated house, we'll give you the full story on weatherstripping and caulking.

There is advice about heating the inside of your home, too: why a fireplace sounds cozier and warmer than it really is; how raising the humidity can permit you to lower the thermostat; what Uncle Sam will do for you if you try solar heat. If you have plants, you can find out how best to get them through the winter, and how they can help make your winter surroundings warmer and more inviting.

And because we all have to go outside some time, we'll tell you everything you need to know about dressing to meet the elements: how to bundle up your children so they will stay warm and dry but still be mobile; how to buy the best down jacket available for what you want to spend; how to layer your clothes for maximum insulation.

And then there is the family car. If you plan to leave the job of winterizing to your local service station, we'll give you a quick lesson in how to be a smart customer; that way you can tell them what needs doing (and you'll understand why). And if you plan to do some of the work yourself, we'll tell you how. You'll also find tips on safe winter driving: how to avoid road emergencies, how to deal with them when they occur.

Keeping warm also means keeping the body well fueled, come storms, colds and flus, and fancy dinner parties. Cooking for those eventualities and more are all part of the story.

So even if the snow is piling up outside your door and the wind is whistling through the eaves, stoke up the woodburning stove, draw up an easy chair—or nestle under a down comforter—and investigate the possibilities for keeping warm this winter, and many winters to come.

Preface: What's Happening to the Weather?

In financially battered New York, businesses lost millions of dollars as snowstorm after snowstorm, including the city's worst blizzard in a decade, lashed the nerve center of American commerce. January 1978 was the wettest beginning of a new year on record for the New York metropolitan area.

Down south, Tennessee was blanketed by three major snowstorms during an eleven-day period and neighboring Kentucky had to declare a state of emergency.

In Ohio, twenty-seven counties were declared disaster areas when two blizzards in a row blocked roads and closed offices. In Indianapolis, Indiana, and in the state of Massachusetts, meanwhile, snow-removal budgets shot into the red.

On the West Coast, a two-year drought came to an end in mid-December, as jubilant farmers breathed a sigh of relief. But their joy washed out as storm after storm saturated California, causing devastating floods and mud slides. By early March, more than 67.5 centimeters (27 inches) of rain had fallen on the city of Los Angeles; it was the city's heaviest rainfall since 1889-90, and nearly twice the *yearly* average of 35 cm (14") for this normally sunny city.

Coming after the frigid winter of 1977, was the winter of '78 another installment in a healthy weather pattern turned malignant? How did it stack up against the previous winter, a winter many meteorologists call the coldest of the century?

The winter of 1976-77 had begun with a cold autumn, as a north-south ridge of high-pressure air built up over the Rocky Mountains. This ridge had created a block that prevented the prevailing westerlies from carrying the warm air from the Pacific Ocean across the continent to the eastern half of the United States. Instead, the warm air had been diverted up to Alaska and the Arctic.

At the same time, a trough of low-pressure air had built up over the eastern part of the United States. This low-pressure trough pulled in frigid Arctic air and carried it down to the South, which is not accustomed to extremely cold autumns and winters. This weather pattern produced the coldest September through November months in the northeastern part of the United States in over ninety years.

The cold autumn of 1976 was followed by a mild December; but then, at the beginning of January, the cold set in again. During January 1977, average daily temperatures were 10°C (18°F) below normal for most of the East Coast, the Midwest and the South, and

3.3°- 5.5°C (10° - 15°F) below normal in Florida and Texas. Until mid-February, the states east of the Mississippi remained in the grip of the extreme cold.

But the winter of 1977-78 was not the same story. According to Dr. Donald L. Gilman, head of the Long Range Prediction Group of the National Weather Service, winter came fairly late to most of the country. In the Great Plains, December was cold, but in the rest of the country, the cold temperatures did not hit until January and February. Overall, the temperatures did not reach the extremes of 1976-77, but the cold was more widespread than in the previous year, extending straight from the Rockies to the Atlantic Coast, instead of being confined to the area east of the Mississippi.

Whereas the story in 1976-77 was record cold, the winter of 1977-78 brought record precipitation. In the Northeast, it came in the form of snow. During January, approximately two dozen weather stations reported record snowfalls.

Did the same type of weather system produce the two record winters? Were these abnormal weather patterns? According to Dr. Gilman, there was nothing unusual about the weather system that produced the frigid winter of '76-'77. The high-pressure ridge over the Rockies that combined with a low-pressure trough over the eastern states is one of many common weather patterns. What was unusual about that winter was that the pattern stayed in place for four months. Usually, such a pattern is replaced by one that moderates the temperature. Why did that particular pattern stay in place? "No one has offered a convincing cause-and-effect explanation for the persistence of the pattern," says Dr. Gilman.

On the other hand, the weather pattern of the winter of 1977-78 *was* unusual, according to Dr. Gilman. The previous year, the Pacific Ocean westerlies had not reached the west coast of the United States. There was a split pattern of westerlies in '77-'78, with a northern belt that carried them up to Alaska and then back down into the eastern states, and a southern belt that slammed into California (producing the torrential rains) then proceeded strongly eastward. It was the southern-belt pattern that departed from the norm, says Dr. Gilman. As these winds spread across the southwest and into the Gulf states, they picked up moisture and power from the Gulf of Mexico.

The harsh and wet winters of 1976-77 and 1977-78 have probably pushed warmer memories from our minds, as we recall flooded basements, snow-blocked highways, and job cutbacks. But do you remember the not-too-distant winter of 1974-75? Some experts had forecast that that winter would be exceptionally cold, but instead it was unusually mild. For instance, in Boston, it was the warmest winter in twenty-two years, while it was Moscow's warmest winter in 230 years.

What does it mean when an "abnormally" warm winter such as 1974-75 is followed a few years later by such "abnormally" cold and wet winters as 1976-77 and 1977-78? Or was the warm winter of '74-'75 the end of a benign climate trend, and the cold winter of '76-'77 the signal for a period of intense cold and snow? Are we perhaps on the verge of another ice age? What is normal?

First of all, when a weather forecaster says "normal," he or she is basing the norm on the characteristics of a period lasting only thirty years. This is a convention adapted by meteorologists, and these thirty-year periods are updated every five years.

The main problem faced by climate experts in examining a season such as the winter of 1976-77 is to distinguish between year-to-year fluctuations and changes in climatic patterns that spread over decades, centures, and even millennia.

Around the middle of this century, scientists discovered that there had apparently been a global warming trend during the first half of the century. There is still controversy as to the exact dates of this warming trend and its global extent. Some scientists say it started around 1890 and ended around 1940; others put the starting date as late as 1918 and the end in the early 1950s. The warmest year in the Northern Hemisphere until the late 1930s was 1915, while 1918 was the coldest of any year in the first three-quarters of the twentieth century. In any case, it is generally accepted that the warming trend did occur and that it produced what may have been the warmest weather the world has known since a hot spell of about 5,000 to 7,000 years ago, following the end of the last ice age.

The recognition of this warming trend prompted scientists to ask important questions. Was the warmer period evidence that the worldwide climate is undergoing a significant change, or is it fluctuating around a long-term "normal" (a "normal" that covers a period of a century or more rather than thirty years)? Moreover, if this is a significant change rather than a random fluctuation, is it caused by human activities? These are still crucial questions, and even now there are no definitive answers. But scientists are continually learning more about past climates and trends.

Increased knowledge of climates during this millennium reveals other distinct cycles, periods of cooling and of warming. Going back almost one thousand years, there apparently was a warming trend in Europe around the time of the Crusades (A.D. 1100 to 1300). Citing archeological data in the United States, Dr. Reid A. Bryson, Director of the Institute for Environmental Studies at the University of Wisconsin, points to a similar warming trend in North America between about 1000 and 1200, followed by a cooling trend roughly during the period 1200 to 1400. We also know that in both Europe and North America, average temperatures dropped enough to produce what is known as the "Little Ice Age," lasting from about 1430 to 1850, with the lowest temperatures occurring during the reign of King Louis XIV of France (1638-1715), who clearly was known as the Sun King for very different reasons. During this period, according to Dr. Bryson, the temperature in the North Atlantic dropped one-sixth of the way to a full-fledged ice age. The population of Iceland declined as a series of famines were experienced, sea ice expanded, European winters were very cold, and mid-summer frosts were common in the north central United States. By comparing current photographs of certain Swiss villages with prints dating from this period, it is possible to see how glaciers crept closer to the villages, then later retreated.

This cooling trend ended some time in the early nineteenth century. It was followed by the warming trend at the beginning of the twentieth century. There is some disagreement about how long that trend lasted. Dr. Bryson says the warming trend ended around 1945. Dr. Hurd C. Willet of the Solar Climatic Research Institute in Cambridge, Massachusetts, a noted expert on the relationship between climate changes and solar cycles, and Dr. John J. Prohaska, also of the Solar Climatic Research Institute, say their data indicate the height of the warming trend was in the early 1950s, with 1953 as the year of peak warmth. According to Drs. Bryson, Willet, Prohaska, and other climate specialists, since the 1950s average temperatures have fallen almost as much as they rose during the first half of the century, by about 0.5 to 1.0°C (1 to 2°F). This has been enough of a change, says Bryson, to have shortened the growing season in England by two weeks a year.

Before we become unduly alarmed by this current cooling trend, we should examine our knowledge of weather cycles not on a scale of centuries, but on a scale of millennia.

We now know from geological evidence that on a global basis, the climates of the past were predominantly warmer than today's. We also know that beginning two million years ago, with the commencement of the relatively cold Pleistocene Epoch, there were a series of ice ages during which glaciers extended into the middle latitudes in both the northern and (according to many scientists) southern hemispheres. Each of these ice ages lasted about 75,000 years, and was followed by an interglacial period with milder temperatures. According to the classic timetable, there were four great ice ages in the last million years. New evidence, however, shows there were more—perhaps seven complete changes from glacial to interglacial conditions. The last ice age was at its peak about 18,000 years ago. Many scientists believe global climates are still on the average a few degrees colder than they were during most of earth's history.

Where does this leave us? What does the present cooling trend mean in light of the particularly cold winters of 1976-77 and 1977-78? How does such a cooling trend fit within the greater context of a global climate that supposedly has a potential for getting even warmer?

Unfortunately, there is no consensus among the top scientists as to what is happening with the weather, either in terms of thirty-year trends or longer ones. But there are many theories.

Many people, both scientists and laypeople, fear that human activities have upset the earth's climate for all time. They are particularly concerned about the effects of carbon dioxide pollution (from the burning of fossil fuels such as oil, natural gas, and coal) and of gases from aerosol spray cans.

Scientists generally agree that the introduction of large amounts of carbon dioxide into the atmosphere can alter the climate. But there is no agreement on what constitutes a large enough amount to pollute the atmosphere, and what the specific effects of this pollution would be. According to one theory, carbon dioxide in the atmosphere creates a so-called greenhouse effect. It allows solar energy to reach the earth, but prevents that energy, in the form of heat, from leaving the earth again. By absorbing the sun's radiant heat, the carbon dioxide thus contributes to an overall warming of the climate.

Some scientists initially claimed that carbon dioxide pollution was responsible for the warming trend of the first half of this century. They based their assertion on the fact that the rise of modern technology, which increased the amount of carbon dioxide in the atmosphere, coincided with the years of the warming trend. But now we know that trend ended in the 1950s, although there is no reason to think that the amount of carbon dioxide in the atmosphere has decreased since then. Scientists who believe in the greenhouse effect have countered with a theory that asserts that something is interferring with its mechanism. The culprit: dust and smoke, of either industrial or volcanic origin, or both.

The role of dust in the atmosphere is very complex. It is true that increasing the amount of dust in the atmosphere would cut off sunlight, thus preventing some of the sun's warmth from reaching the earth. This could theoretically counteract the heat effect of carbon dioxide pollution. Dust in the atmosphere, however, would itself be heated by solar radiation, thus warming the atmosphere at the levels on which it circulates. Scientists claim that dust may also absorb radiation attempting to leave the earth, thus intensifying rather than inhibiting the greenhouse effect.

Our atmosphere has a layer of ozone, a form of oxygen, concentrated between fifteen and thirty miles above the earth's surface. This ozone layer acts as a blanket, protecting the earth from some of the sun's potentially harmful ultraviolet radiation. Many scientists have claimed that the gases used to propel shaving cream, deodorants, household cleansers, and other products packaged in spray cans have the ability to destroy the ozone layer. Loss of this layer would result in less protection from the sun's radiation, and would almost certainly warm the earth's climate.

Scientists have created countless models for these and other theories of man-made pollution. To date, none has been accepted by the majority of the scientific community as adequate explanation for either the warming trend of the first half of the century or of the subsequent cooling trend. In fact, two noted climate experts, Dr. Helmut E. Landsberg, Director of the Institute of Fluid Dynamics at the University of Maryland, and Dr. Murray Mitchell, of the Environmental Data Service of the National Oceanic and Atmospheric Administration (NOAA) in Boulder, Colorado, see no conclusive evidence at all of the effect of man-made pollution on worldwide climate. Both men agree that too little is still known about the effects of air pollution to assess its role in climate change. Referring to the cooling trend since the 1950s and the previous half-century of warming, Dr. Mitchell says, "There is still no widely accepted explanation for either one."

A theory frequently advanced to explain both short-term and long-term variations in the earth's climate is that variations exist in the intensity of solar energy reaching the earth. One factor that might cause variations in the level of solar energy is variation in sunspot cycles. According to this theory, an increase in sunspot activity projects more solar particles, and thus solar radiation, toward the earth. Scientists who believe in the effects of sunspot cycles say droughts and hot spells on earth occur in step with sunspot cycles of eleven years, twenty to twenty-two years, and even longer, perhaps as much as several hundred years.

One proponent of the theory linking sunspot cycles and climate is Dr. John A. Eddy of the National Center for Atmospheric Research. Citing tree-ring records as evidence of sunspot cycles, Dr. Eddy asserts that whenever the sun seems to lose sunspots on a major scale, the earth goes through very cold spells, such as the "Little Ice Age" that reached its peak in the seventeenth century. Conversely, during periods of increased sunspot activity, the earth goes through a period of warming. Dr. Eddy does not, however, think sunspot cycles are either the exclusive or even the most important influence on the climate.

Dr. Hurd C. Willet is one of the leading proponents of the solar climatic cycle theory (i.e., changes in climate produced by sunspot cycles). He believes recent changes in climate as well as past changes can best be explained by solar patterns. Dr. Willet, together with Dr. Prohaska, notes that over the past three hundred years, prolonged periods of very low sunspot activity have been cold and wet in middle latitudes (most of the United States), while prolonged periods of markedly rising sunspot activity have coincided with periods of warming in middle latitudes, to the highest levels on record. Drs. Willet and Prohaska correlate the sunspot minimum of 1650-1700 with the peak of the "Little Ice Age" in North America and Europe, the sunspot maximum of 1778 with an extremely warm period in the third quarter of the eighteenth century, and the sunspot maximum of 1957-58 (the highest sunspot maximum in five hundred years) with the warming trend of the second quarter of the twentieth century. They claim the cooling period of 1954-1973 was a period of strong decrease in sunspot activity. And they further note that the temporary surcease of the cooling trend during 1973-1975 and the first half of 1976 coincided with a "sharp increase in geomagnetic activity [due to sunspots] to quite high levels in 1974." After the spring of 1976, however, it fell sharply again, and with this fall there was a return to predominantly cold weather.

Drs. Willet and Prohaska base their theories on two sunspot cycles of one hundred years each, separated by a less active and less warm sunspot cycle of eighty years. They predict that "during the next twenty-five years (analogous to the first quarter of the nineteenth century in the past eighty-year cycle), the cooling trend will reassert itself very strongly." In other words, they are saying, batten down the hatches folks, there's twenty-five years of cold weather ahead!

But does this cooling trend mean we are headed for another ice age? Emphatically no, says Dr. Willet. Although he sees future warming and cooling trends in line with sunspot cycles, he does not think another ice age is likely for at least 10,000 years, and probably not for another 30,000 years. Following the cooling trend that he predicts will last until the end of this century, Dr. Willet sees an abrupt return to markedly warmer weather, lasting only ten years, followed again by a return to cooler temperatures.

The solar climatic cycle theory is not currently the most popular one in the scientific community. But there is another solar variation theory, used to explain only long-term fluctuations in the weather, that has gained greater credibility in recent years.

In the 1920s, Yugoslav scientist Milutin Milankovitch proposed that ice ages, the most dramatic results of climatic changes, were triggered by cyclical variations in the earth's spin axis and orbit. Over a period of tens of thousands of years, the shape of the earth's orbit around the sun changes shape. At times it is almost circular, with little variation in the distance between the earth and sun from one time of year to the next. Because of gravitational pull from other planets, however, the earth's orbit changes shape. When it does, the earth's distance from the sun may vary each year by several million miles. This variation in the shape of the earth's orbit is thought to occur in cycles of approximately 93,000 years.

The tilt of the earth's spin axis in relation to its orbit also varies, over a period of approximately 41,000 years, from 22.0 to 24.5 degrees. And the direction of the earth's axis, in relation to the stars, changes once approximately every 26,000 years. According to Milankovitch, the combined effects of these variations accounts for the advent and passage of ice ages. Although for many years the changes implied by Milankovitch's theory were considered to be too minute to trigger ice ages, Milankovitch's theories have gained better repute in the last few years than at any other time in the past twenty-five years, according to Dr. Jerome Spar of the City College of New York. Milankovitch's theory states that as long as the earth's tilt axis and solar orbit remain in the same approximate position as they are at present, the advent of another ice age is very unlikely.

When it comes to the firing line, very few climatologists believe another ice age is imminent. But when it comes to predicting the weather for the next decade, the next year, and even the next season, very few climatologists or meteorologists will go out on a limb. The only thing most of them will probably agree on is that the climate is undependable: there is just no such thing as a constant climate.

What about the U.S. government? They have a National Weather Service. Can't the National Weather Service make long-range predictions? They even have a long-range prediction group; what does it do? In this case, long-range only means the next month and the next season. Each month, at the very end of the month, the long-range prediction center tries to predict the weather for the coming month. And a few days before the next season, it tries to predict the weather for the coming season. But both the monthly and seasonal predictions are still done on an experimental basis, with varying degrees of success. So far, at least, they have been more accurate about temperatures than precipitation. The time when anyone will be able to predict the next month's, season's, or year's weather with any precision is still in the far future.

What does this mean for the coming winter? No one can say for sure, but the best thing you can do is be prepared for any cold, wind, and snow that comes your way.

Insulation:Wrapping up the Home for Winter

Winter has descended. Snow is gently falling, swept into drifts outside a casement window; a sweetly scented pile of logs burns in a bright blaze in the fireplace; a dog and cat lie contentedly on the hearth; bread bakes in the oven; a woman rocks and knits, her husband cheerfully works on a piece of wood and their children play nearby. With a sigh of nostalgia, most of us can conjure up such Laura Ingalls Wilder pictures of a winter's day in the country in the last century. However, the reality of rural life was something quite different. Those casement windows were drafty; the house was often poorly insulated. The fireplace may have smoked and, in any event, was a terrible heat waster. There was little time for sitting around. The woman was often faced with an iced-over kitchen and frozen larder, while the man had to contend with the myriad demands of livestock and land, and the children had their own unending chores. The primary challenge of winter was keeping warm, and failure meant the loss of livestock and maybe life.

Today things aren't so dramatic, but heating our homes is still the foremost task of the winter months. The largest single expense in running a house is keeping it heated (and if you live in a warm climate, cooled). Over the past five years, these energy costs have tripled—and they will continue to rise. But even if you can do nothing to control the basic costs, you can work to reduce energy consumption. That doesn't mean you won't keep warm; it does mean keeping heat loss to a minimum.

This may require new insulation, an upfront investment which will pay for itself in fuel savings. And with recent government interest in conserving fuel, you may even get some assistance, both financial and informational, in making energy-saving home improvements.

Whether your primary interest is keeping warm for the least amount of money, or just keeping warm period, the first step is the same: wrap up! Wrap up your house the way you wrap up yourself and it will keep warm. When it comes time to heat your house, you don't want to heat the whole outdoors. The tighter your house is, the less precious heat will escape and the less cold air will enter.

Let's look at what this means in specific, practical terms.

Heat and How It Is Lost

Any discussion of keeping warm is also about keeping away cold. Cold itself is not a measurable quantity; it is defined simply as the

Winterizing Help from the Government

If you own your home and need help weatherizing it, you may be able to get that help from the government. Ever since the energy crisis, several U.S. government agencies have taken an active interest in warming houses more efficiently. Some programs are already underway and many more will begin in the near future. If these agencies don't have the information and the know-how you need, they can put you in touch with people who do. Some of these agencies can even front the money you need to do the work and maybe the manpower as well.

U.S. Department of Housing and Urban Development (HUD)

HUD is interested in raising all housing above the national minimum standards of comfort and efficiency. At the moment, all they give out directly is information, but they are also the parent agency of the *Federal Housing Administration.* The FHA guarantees home-improvement loans which participating banks provide homeowners at reasonable rates. HUD has ten regional offices, each of which has a public information officer who knows what is available in your particular region and can steer you to the right place.

Region   Address

I HUD Public Information Office
Room 800, John F. Kennedy
Federal Building
Boston, MA 02203
(617) 223-4066

II HUD Public Information Office
26 Federal Plaza, Room 3541
New York, NY 10007
(212) 264-8068

III HUD Public Information Office
Curtis Building
6th and Walnut Streets
Philadelphia, PA 19106
(215) 597-2560

IV HUD Public Information Office
Room 211, Pershing Point Plaza
1371 Peachtree Street, N.E.
Atlanta, GA 30309
(404) 526-5585

V HUD Public Information Office
300 South Wacker Drive
Chicago, IL 60606
(312) 353-5680

VI HUD Public Information Office
Room 14C2, Earle Cabell Federal
Building, U.S. Courthouse
1100 Commerce Street
Dallas, TX 75242
(214) 749-7401

VII HUD Public Information Office
Federal Office Building, Room 300
911 Walnut Street
Kansas City, MO 64106
(816) 374-2661

VIII HUD Public Information Office
Executive Tower
1405 Curtis Street
Denver, CO 80202
(303) 837-4513

IX HUD Public Information Office
450 Golden Gate Avenue
P.O. Box 36003
San Francisco, CA 94102
(415) 556-4752

X HUD Public Information Office
Arcade Plaza Building
1321 Second Avenue
Seattle, WA 98101
(206) 442-5414

Veterans Administration (VA)

If you are a veteran, you can get a GI loan to pay for the purchase or improvement of a house. The VA does not make these loans directly, but can tell you if you are eligible, and if so, for how much money and to which banks to go to in your area. There are many VA offices in each state. If you don't know where the one in your area is, look in the government listings in the telephone directory. There is a toll-free number to call in each state.

U.S. Department of Energy (DOE)

For the present time, most of this department's programs and credits for energy conservation are still on the drawing board, but it can still provide you with useful information and steer you to other agencies in your area. In addition, there is one program that is underway, a weatherization program for low-income families. To find out if you qualify (or to keep in touch with pending programs), call or write the DOE public information office for your region.

Region   Address

I DOE Public Information Office
Analex Building, Room 700
150 Causeway Street
Boston, MA 02114
(617) 223-3701

II DOE Public Information Office
26 Federal Plaza, Room 3206
New York, NY 10007
(212) 264-1021

III DOE Public Information Office
1421 Cherry Street, Room 1001
Philadelphia, PA 19102
(215) 597-3890

IV DOE Public Information Office
1655 Peachtree Street, N.E.
8th Floor
Atlanta, GA 30309
(404) 526-2837

V DOE Public Information Office
175 West Jackson Boulevard
Room A-333
Chicago, IL 60604
(312) 353-8420

VI DOE Public Information Office
P.O. Box 35228
2626 West Mockingbird Lane
Dallas, TX 75235
(214) 749-7345

VII DOE Public Information Office
Twelve Grand Building
P.O. Box 2208
112 East 12th Street
Kansas City, MO 64142
(816) 374-2061

VIII DOE Public Information Office
P.O. Box 26247, Belmar Branch
1075 South Yukon Street
Lakewood, CO 80226
(303) 234-2420

IX DOE Public Information Office
111 Pine Street
Third Floor
San Francisco, CA 94111
(415) 556-7216

X DOE Public Information Office
1992 Federal Building
915 Second Avenue
Seattle, WA 98174
(206) 442-7280

Community Services Administration (CSA)

This is the new name of the former Office of Economic Opportunity (OEO). It has joined forces with the DOE to provide manpower as well as materials for weatherizing housing in low-income areas. If you want to find out if there is a CSA program in your area, check the government listings in your local phone book—there are more than 865 local agencies. If you don't find one, write to the Office of Public Affairs, Community Services Administration, 1200 19th Street, N.W., Washington, DC 20505, or call (202) 254-5150. You can also write or call your regional DOE office.

Farmers Home Administration (FmHA)

As part of the energy-conservation drive, the FmHA is offering loans to people who want to install nonfossil energy systems in any residential building in a rural area. Rural areas are defined as places with populations of 10,000 or less, but if you live in a town of 10,000 to 20,000 people and there is a serious lack of mortgage money, you may qualify. There is an FmHA office for each rural county.

State Energy Agencies

Every state now has an energy department. The information and other assistance that you get from them may be more specifically relevant to your conditions than that available from the federal agencies, and may be coordinated with federal programs. Some states are more actively involved in energy conservation than others; to find out about yours, write or call your state office.

absence of heat. Heat can be generated, moved, measured and lost. When it is lost, you may feel cold, but physicists insist that it's the absence of heat you are feeling, not the presence of cold. This is not mere semantic juggling; we want to make the point now so you will understand why our discussion focuses on moving heat rather than on excluding cold.

The key to the whole "tight house" and insulation issue is the physical fact that heat moves from one place to another, depending on a number of factors; your object is to control the movement.

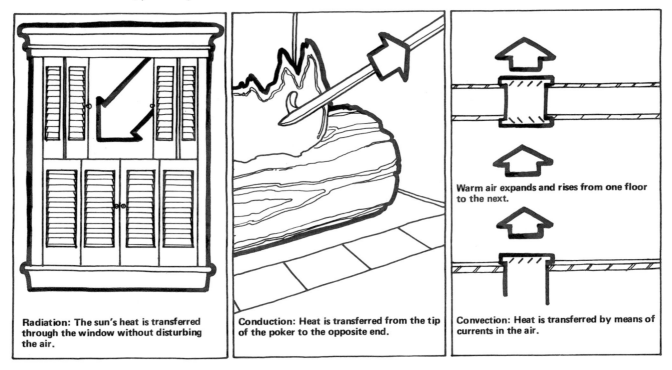

Radiation: The sun's heat is transferred through the window without disturbing the air.

Conduction: Heat is transferred from the tip of the poker to the opposite end.

Convection: Heat is transferred by means of currents in the air.

Warm air expands and rises from one floor to the next.

Glossary of Terms

Heat Measurement

British Thermal Unit (BTU). This is one of the standard units by which heat is measured. One BTU equals the amount of heat required to raise the temperature of water one degree Fahrenheit. Heat from burning fuels is usually measured in BTUs.

Calorie. There are actually two types of calories: the greater calorie, or kilocalorie, and the lesser calorie. Both are standard units for measuring heat transference. The lesser calorie represents the amount of heat required to raise the temperature of 1 gram of water 1 degree centigrade. 1000 lesser calories equal 1 kilocalorie. A kilocalorie is the amount of heat required to raise the temperature of 1 kilogram of water one degree centigrade. Kilocalories, called simply "calories," are commonly used to describe the heat value of foods.

Therm. This is another metric unit of heat measurement. Like the lesser calories, it represents the amount of heat required to raise the temperature of water 1 degree centigrade. It has been used to measure gas heat and to determine charges for gas used.

Heat Transference

Conduction. This is the flow of heat through a stationary object, from a hot part to a cooler part. For example, when you leave a metal spoon in a hot cup of coffee, the heat moves very quickly up the handle,

which becomes too hot to touch. Some materials are better conductors than others, air and wood make poor conductors, while copper and silver are excellent.

Convection: When a fluid (gas or liquid) is heated and flows to a cold body, the heat transferred to the cold substances moves by convection. For example, a fire built in a space heater in the center of the room will heat the air above it. This hot air rises, heating the air above and spreading out to the walls of the room. As the air cools it will sink to the floor and be drawn back to the heater to begin the process all over again. This is heat by convection, and the air currents produced are called convection currents.

Radiation. Heat which flows from a hot body to a cold one without passing through a solid or liquid moves by radiation. For example, the sun's heat moves to the earth by radiation..

Insulating Values

K. The K-factor represents conductivity. Specifically, the K-factor is the amount of heat which will pass through a piece of material measuring one foot by one foot by one inch per hour per degree Fahrenheit of temperature difference from one side to the other of the material. For example, if the type of insulation you are considering has a K-factor of 2, this means that 2 BTUs will be lost every hour for every degree of temperature difference from inside the house to outside. If it is 15° colder outside, then 30

BTUs are lost. Thus the lower the K-factor, the better the insulation.

C-value. The C-value also designates the conductivity of a given material. C-values are assigned to materials based on different thicknesses. Like the K-factor, it represents the amount of heat lost per hour, per square foot, per degree Fahrenheit of temperature difference between two surfaces. For example, the C-value of an eight-inch hollow concrete block is 1.00; that is, one BTU is lost every hour over each square foot for every degree difference between the inside and colder outside surfaces.

U-factor. This is used to measure the overall heat transferred from air on one side to air on the other side of a combination of materials which make up a wall, ceiling or roof. High U-factors mean high heat loss.

R-value. This is the resistance value of a given combination of materials to heat loss. An R-value assigned to an insulating material refers to the material's resistance when properly installed in a wall, floor or ceiling. The higher the R-value, the lower the heat loss and the better the insulator. Different R-values are recommended for different areas of the house according to which ones are more prone to heat loss.

Heat moves (or is transferred) by radiation, conduction or convection. When heat is transferred from one surface to another without disturbing the air between these surfaces, it is being *radiated.* The sun's heat moving through a window is radiated. When heat is transferred directly through one material to another it is moved by *conduction.* Heat travels from the tip of a metal rod poked in a fire to its end several feet from the fire by conduction. And when heat is transferred from air or water of high density to air or water of lower density, it moves by *convection.* Denser air or water is colder and tends to fall; less dense air or water tends to rise. Hot air rising from a lower floor to the attic moves by convection.

If you consider these three methods of transferring heat, you will soon realize that air is a good medium for radiation and convection, but it is a poor conductor. Because of this, air makes an excellent insulator, or substance that resists heat transfer and therefore heat loss. All materials are insulators to some degree or other, but some are better insulators than others because they resist heat more effectively.

If air is a good insulator, trapped air is even better. Trapped or compressed air pockets resist heat transfer because in order to pass through the trapped air, the heat must constantly shift its mode of travel from conduction to radiation and back. Where air pockets meet, heat is conducted; in between, it must be radiated. This process slows down heat transfer. And the less porous the insulating material, the slower heat can move through it and the more efficient it is at resisting heat transfer.

Selecting Proper Insulation

The R-Factor
Resistance to heat transfer is called the R-factor, and all insulation products are measured in terms of their R-value. The higher the R-value, the more effective the insulator; you should select an insulator based on its R-value, not on its thickness alone.

Heat is not lost evenly throughout the house, and the amount of insulation varies in the different parts of the house. Since hot air rises, the greatest amount of heat—some 25 percent—is lost through the attic and roof of the house. Much less heat is lost through the walls, and still less through the floors. These areas of the house will need different types of insulation. To guide you in selecting the proper insulation, the federal government has set specific standards regarding the amounts of insulation necessary. (The federal recommendations were adjusted upward in 1976 as part of the government's effort to improve energy efficiency and conservation.)

This system for measuring insulation needs was developed from observation of temperature ranges in various climates and calculations of the ideal differentials between indoor and outdoor temperatures. In order to figure out accurately what R-value insulation you require and where, you will need to know certain things about the climate in which you live. And that brings us to the subject of degree days.

Degree Days.
A degree day is a unit of measurement. It is based on the assumption that heat is not needed when the outdoor temperature is 65° F or more, but is needed when outdoor temperatures fall below that. How much below 65° F tells you how much heat is needed. Heat-

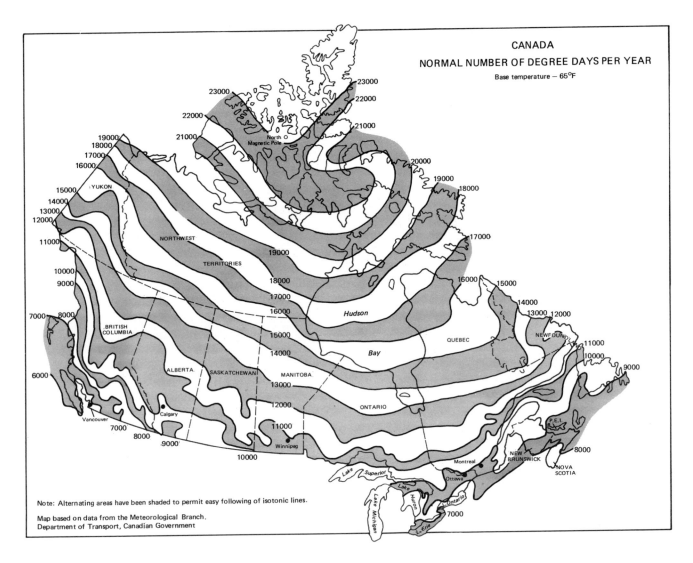

CANADA
NORMAL NUMBER OF DEGREE DAYS PER YEAR
Base temperature — 65°F

Note: Alternating areas have been shaded to permit easy following of isotonic lines.

Map based on data from the Meteorological Branch.
Department of Transport, Canadian Government

ing Degree Days (HDD) are calculated by subtracting the mean temperature for a given day (the highest temperature plus the lowest, divided by two) from 65° F. For example, if the lowest temperature on the day in question was 38° F and the highest was 50° F, the mean is 44° F. Subtract 44 from 65 and you get the HDD for that day, 21. All this figuring becomes useful when compiled over the course of a heating season. The National Weather Service does just that. Look at the map above and you will see how the country has been divided into degree-day zones. If you find your home town, make note of the number and you will be well on your way to figuring out what your residential insulation and heating needs are. The chart on page 16 breaks down the recommended minimum R-values for various areas of the typical single-family frame house in the different climate zones of the United States.

Types of Insulation.

Whether you are insulating a new house or an old one, you will find a wide choice of insulating materials of varying thicknesses and R-values to suit your needs. One specific problem with all insulation, except foam and polystyrene, is its tendency to absorb and hold moisture. To prevent this, a vapor barrier is necessary. Attached vapor barriers are usually made of foil; separate vapor barriers are simply rolls of heavy-duty plastic or polyethylene sheeting. The vapor barrier

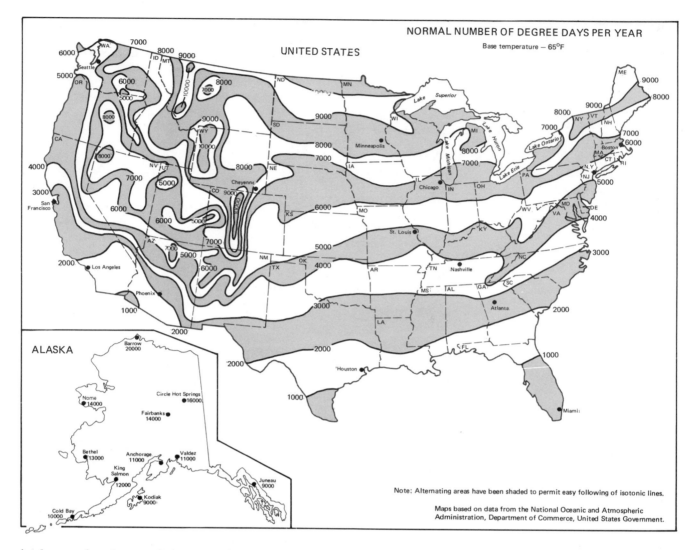

is always placed toward the warm living area of the house.

The various insulation materials now available include the following.

Blanket: Fiberglass or rock wool. Comes in rolls, 16 in. (40 cm) or 24 in. (60 cm) wide, to fit between joists, studs or rafters. Thickness varies from 1 in. (2.5 cm) to 7 in. (17.5 cm). Comes with or without vapor backing, which will have flanges extending from either side for easy nailing or stapling.

Batt: Fiberglass or rock wool. Comes in sections, 4 ft. (1.2m) or 8 ft. (2.4m) long. Widths and thicknesses similar to blankets; comes with or without vapor barrier.

Board: Natural material such as plywood is not recommended. Instead, use a rigid foamed-plastic or fiberglass board. Ranges in size from 8 in. (20 cm) squares to 4 ft. by 12 ft. (1.2 m by 3.6 m) panels. Thickness ranges from ¾ in. (1.8 cm) to 4 in. (10 cm). Moisture resistant, so needs no vapor barrier. Board-foam insulation has a high R-value, but the foamed plastic type is combustible (although most melt rather than burn). Building codes require sheathing of ½-in. (1.3 cm) thick gypsum board over this material on interior surfaces. They are usually installed with a special glue applied directly to the studs.

Poured-in loose fill: Fiberglass, rock wool, cellulosic fiber, vermiculite and perlite. Comes in bags; pour between joists or between walls. Tends to settle in vertical cavities and must be replen-

14   Insulation

ished to be effective. Poured-in insulation needs a separate vapor
barrier, usually polyethylene sheeting, but when used in walls, a
couple of coats of moisture-resistant paint on the interior walls will do.

Blown-in loose fill: Fiberglass, rock wool, cellulosic fiber.
Can be used in the same places as poured-in fill, and needs a separate
vapor barrier; but this type of insulation should be installed by a professional contractor with the necessary equipment. Holes are made in
the exterior walls between the studs and fairly high up, and the fill is
introduced under pressure into the cavities.

Types of insulation

Batts: 4 or 8 feet long, 16 or 24 inches wide.

Blankets: rolls, 16 or 24 inches wide. Both with or without vapor barrier.

Loose fill: poured-in or blown-in.

Rigid board: 8-inch squares or 4 x 8 foot panels.

Blown-in foam or loose fill.

Farenheit-Celsius Conversions

Beginning in June 1978 the National Weather Service, in accordance with the plan for conversion to the metric system, stopped using Farenheit as the standard measurement of temperature and adopted Celsius. If that makes it difficult for you to figure out how cold is cold, the accompanying chart will provide a key. If you want to do your own conversions, the formula for changing Farenheit to Celsius is

$C = 5/9 (F - 32)$

Farenheit can be found as follows:

$F = 9/5 C + 32$

If you are not mathematically inclined and are happy with a "sense" of how cold it is, remember this little ditty:

30 is hot	(86°F)
20 is pleasing	(68°F)
10 is not	(50°F)
0 is freezing	(32°F)

Farenheit to Celsius		Celsius to Farenheit	
-25	-31.7	-30	-22
-20	-28.9	-25	-13
-15	-26.1	-20	- 4
-10	-23.3	-15	5
- 5	-20.5	-10	14
0	-17.8	- 5	23
5	-15.0	0	32
10	-12.2	5	41
15	- 9.4	10	50
20	- 6.7	15	59
25	- 3.9	20	68
30	- 1.1	25	77
32	0	30	86
35	1.7	35	95
40	4.4	37	98.6
45	7.2	40	104
50	10	45	113
55	12.8	50	122
60	15.6	55	131
65	18.3	60	140
70	21.1	65	149
75	23.9	70	158
80	26.7	75	167
85	29.4	80	176
90	32.2	85	185
95	35.0	90	194
98.6	37.0	95	203
100	37.8	100	212
105	40.6		
110	43.3		
115	46.1		
120	48.9		

Foam: Also called "arctic foam." Ureaformaldehyde or urethane. Blown or sprayed in under pressure. Must be installed by a professional. No vapor barrier necessary.

Blown-in foam or loose fill cannot be used in a wall in which there is already some insulation material. In that case, you will have to resort to adding insulation to the exterior of the house and using wall panels inside wherever you can, unless you're willing to take all the walls apart and start from scratch.

If you are planning to hire a contractor to insulate with blown-in foam or fill, make sure you agree on the terms before the job begins: type of insulation needed, amount of insulation and method of installation. You should get a written contract and it is best to check on the job from time to time to be sure everything is going smoothly. It's also a good idea to select a contractor either through the recommendation of someone you trust or by shopping around and getting estimates from a few firms.

In choosing the insulation you'll use, remember that batts and blankets are easy to handle but that poured-in insulation, although fine to work with in the attic, will have to be installed in walls by a professional, as will blown-in foam or fill. Once you've decided on the type, you'll have to figure out how much to use. It is fairly easy to measure the insulation you already have or will need. If you can get into your attic and basement or crawl space, take a ruler and flashlight and measure the depth of the joists and the width between them. Be sure to measure the thickness of any existing insulation at the same time to see whether it complies with government recommendations. If not, you'll want to add to it. Walls are more difficult to get a reading on but you can remove a light switch, turn off the current, and insert a ruler to see if you meet resistance (there's something in there!) or to see how deep the cavity is to the outer wall.

Once you have these measurements, figure out the overall area by multiplying the width by the length of a floor or ceiling, the length

---

Heating Zones and Recommended R-Values

Use this chart of recommended R-values to determine the best insulation for your home.

Zone	Wall	Ceiling	Floor
I	11	19	11
II	13	36	11
III	19	26	13
IV	19	30	19
V	19	33	22
VI	19	38	22

The Federal Energy Administration and the United States Bureau of Standards have established six winter heating zones in the United States. From south to north—warmest to coldest—they are:

Zone I: Florida, the southern parts of the Gulf states, south Texas, southern California and southwestern Arizona.

Zone II: From Zone I north to Virginia and stretching west to include parts of California, Arizona, New Mexico, Texas and Oklahoma.

Zone III: From Zone II north to southern New York, southern Connecticut and west to the Rockies, then looping south to include parts of New Mexico, Arizona, Nevada, California and the Pacific Northwest.

Zone IV: Most of New England, New York, the Rockies and the Northwestern states.

Zone V: Northern Maine, New Hampshire, Vermont, the peaks of the Rockies, North Dakota, most of Minnesota, part of Wisconsin, Michigan, and southern Alaska.

Zone VI: Alaska

The Government has set R-value recommendations for insulating the different areas of homes within these zones. If you are unsure of the zone in which you live, contact the Federal Energy Administration or the Cooperative Extension Service in your area.

by the height of a wall. Take into account the space used by the studs or joists; you'll actually need less than the overall area. If the joists or studs are 16 in. (40 cm) apart, you'll need .9 times the area; if they are 24 in. (60 cm) apart, .94 times the area. To calculate the amount of insulation in a crawl space, crawl in and measure the height of the wall plus the "header" (that space between the subfloor and the wall, which is the same width as a joist); over walls which run parallel to the joists, the header is called a box joist. Add 2 or 3 feet (60 or 90 cm) to the measurement to allow the insulation to bleed out onto the floor. Then calculate the perimeter of the space and multiply by the width of the batts or blankets you will use.

Do-It-Yourself Projects

It is, of course, much easier to insulate a house as it's being built than an old one. If your house isn't brand new, chances are it will need some additional insulation—especially to catch up with the newest government recommendations. But for a new home or old, check out the tax write-offs which have been set up by the federal government as incentives to save energy; they may just save you money too.

Attics

The unfinished floor in the attic lets heat escape in the winter and allows unwelcome heat to enter in the summer. Yet it is simple to avoid this by insulating between the joists. You can use batts, blankets or loose fill with vapor barriers, or a combination of the three. All are easy to handle for the do-it-yourselfer.

Before you begin, be certain that there are no roof leaks. If necessary, make repairs. Then be sure you have adequate ventilation, not only for your work, but also to avoid condensation once the insulation is installed. For every 300 sq. ft. (28 sq. m) of attic floor, you'll need 1 sq. ft. (930 sq. cm) of vent. Half of the vent area should be in the eave or soffit to encourage air flow up and out the rest of the vents. You can install ready-made gable, eave or soffit vents yourself; select ones with louvers, which can be closed in case of heavy rain.

You will also need lighting to see what you are doing; a clamp-on fixture with a long extension cord leading down to the nearest outlet will do.

A temporary moveable floor is another necessity. Existing ceilings usually cannot support any heavy weight and it is fairly impossible to balance yourself on the joists for any length of time. If you tire you may find yourself making an unexpected and swift descent to the floor below. Planks can be set across several joists and will support you, and your tools and the insulation.

In calculating the amount of insulation you'll need, don't neglect to include some to cover the access door. If your R-factor indicates that you will need two layers of batt or blanket insulation, you will need to lay the second layer at right angles to the first. This layer can be held in place by adding supports at right angles to the joists at intervals the width of the insulation you are using.

The first layer is placed between the joists with the vapor barrier facing downward, toward the living area. If you are using batts or blankets without an attached vapor barrier, first cut sheets or strips of polyethlyene a few inches (5 or 6 cm) wider than the spaces between the joists. Fit the sheets between, staple them to the joists and then carry on with the insulation.

Each batt or blanket should be set between the joists so that it does not press against the eaves; this would block the flow of air through the attic. Work from the eaves toward the center, cutting pieces with a serrated knife or dull handsaw to fit snugly. Where the insulation meets cross braces, it must be cut to fit snugly above and below each brace. Insulation must terminate at least 3 in. (7.5 cm) away from any light fixture set into the floor, but should fit closely around any chimney, with a layer of heavy-duty, noncombustible material—preferably foil—between it and the chimney.

If you already have some insulation in your attic but need more, add another layer of batting or blanket insulation. If the new insulation has a vapor barrier, slash it with a knife. A second vapor barrier would only trap moisture and cause trouble.

If your attic has a finished floor, you can still insulate between the rafters, as well as the end walls. Be sure that the vapor barriers of the batts or blankets face inward, toward you. Staple the flanges of the vapor barrier along the rafters, leaving a little air channel at the peak of the roof.

Because of ventilation, it is preferable to make a kind of ceiling with the batts or blankets below the roof peak, which will leave an air space. Span each pair of rafters with a 2 ft.-by-4 ft. (60 cm-by-120 cm) board cut to measure. These are called collar beams and will form the ceiling. Batts or blankets are stapled between these beams and between the rafters as usual. At their junctures, tape the edges together. Do not try to run continuous insulation up, across and down rafters and beams. Not only is it awkward to do, but there will inevitably be heat-escaping gaps.

If the end walls of your attic have studs, install batts or blankets between them, the vapor barrier facing you. If there are no studs, you can make a frame with two-by-fours, cut to measure the space and nailed to the wall at intervals which correspond to the width of your insulation.

If your attic is entirely finished but not insulated, you will have to cut openings in the ceiling and walls to gain access to the rafters and eaves. Install batts or blankets between rafters and joists you can easily reach; pour loose fill between the rafters you cannot get to.

Walls

Underinsulated walls are a problem even for the most determined do-it-yourselfer. Unless you are willing to tear down your interior walls or paneling to bare the studs, you will have little choice other than poured-in or blown-in insulation, both of which are best installed by a professional. If there is existing insulation in the walls, even this is impossible. However, you can still add some extra protection to old walls by using wall paneling indoors (corkboard and the like will add a bit of insulation) and by adding insulation and/or new siding outdoors.

If you are going to replace your interior walls with new paneling or wallboard, it is very simple to install insulation first. Batts or blankets fit snugly between the studs, vapor barrier facing you. If you opt for batts or blankets without a vapor barrier, you can staple polyethylene sheeting over the insulation before you put up the paneling or wallboard.

If you decide to insulate exposed studs with rigid plastic board, you will have to remember to sheathe it with panels of ½-in. (1.3 cm) thick gypsum board before installation to comply with building codes and to make the plastic more fire resistant.

Insulating from the Outside

You decide that your only alternative is to insulate the walls on the outside. It's a big job, beginning with cleaning the old siding and carefully checking for any leaks or damage. Make any repairs.

Proper installation of the new panels is essential. The polystyrene board is moisture repellent and creates an exterior vapor barrier. Since vapor barriers are effective only on the warm side, improper installation of the polystyrene would actually encourage condensation within the walls. To be effective, vapor barriers must make an airtight box around your house. Seal them tightly with the special adhesives and tapes which are recommended, then proceed with new siding.

Sidings in themselves have insulating qualities and many of the newer ones come with an insulation backing that gives them a higher R-factor.

Roof

You can insulate your old roof in much the same way as exterior walls. Check first for roof leaks and damage in old roofing and make necessary repairs. Place a layer of roofing insulation material over your current roof and install new roofing over that. Many of the new shingles, especially the asphalt variety, come in heavier weights and in colors that add to their insulating effect. Choose a dark color if you are in a northern climate beset with extremely harsh winters. The dark color absorbs more heat from the sun and will keep the attic warmer than a light one. However, if your home has little shade in the summer and needs to be cooled at high cost, you might do better with a lighter color for the roofing; this will reflect the sun's heat and keep the house cooler.

Basements and Crawl Spaces

Although heat does tend to rise, a significant amount is also lost through basements and crawl spaces, especially if they are unheated and have earth floors. Heat loss can be minimized by insulation.

First lay down a vapor barrier of polyethelene sheeting on the earth floor to prevent moisture from seeping in and being absorbed by newly installed insulation. Fit each batt or blanket section on the walls between the studs, starting at the top (at the subfloor) and running down the wall and onto the floor. To affix the insulation at the top, cut strips of scrap lumber to measure the space between the studs and use them to nail each section of insulation to the box joist or header. On header walls, you will have to cut slits in the insulation to fit around the joists.

In a regular basement, insulate the walls either with batts or blankets between studs or with rigid board. If you prefer the former and do not have studs, you can construct a frame of two-by-fours along the walls and then fit the insulation, with the vapor barrier facing you, within the frame. You do not need to go all the way to the floor; in an ordinary basement the insulation need only reach 2 ft. (60 cm) below the outside ground line. However, don't forget to insulate the headers and box joist, too.

If you decide on rigid board, you will need a frame as well, but it can be made with thin strips of board known as "furring," which are affixed directly onto the concrete wall with a special heavy-duty mastic. The furring should be spaced vertically along the wall so that the boards will fit exactly within the frames. The boards are also attached to the adhesive mastic. It is worth repeating that for fire protection, panels of ½-in. (1.3 cm) gypsum board must be tacked to the furring strips over the insulation once it is installed.

Both batting and board insulation will provide a backing for paneling or wallboard if you want to convert your basement into a more usable room.

The basement ceiling, between the first floor joists, should be insulated with batts or blankets, the vapor barrier facing upward toward the living area. To hold them in place, staple wire mesh or chicken wire across the joists.

When you are insulating your basement, remember to wrap all exposed ducts with batting or blanket insulation held in place with special duct tape. Exposed pipes require special pipe insulation material and insulation tape to cover the fittings.

Like the attic, both basement and crawl space will need adequate ventilation to prevent condensation or moisture buildup. You should have one vent for every 300 sq. ft. (28 sq. m) of floor area. It is best to install ventilators from the outside close to the corners on two opposite walls, and near the top of your foundation walls. These vents should have louvers so that they can be partially or totally closed if necessary.

Natural Insulators

After you have insulated the indoor space and worked on outdoor insulation where needed, you can add another dimension to the system by letting nature give you a hand. A good place to start is at ground level. Dampness builds up against the foundation walls and moisture can creep into your basement or crawl space through the walls' porous material, especially concrete block. One way to prevent this is earth berming. This simply means that you mound earth up against your house, to a point at least slightly above ground level and higher if you wish. The more earth, the better the insulation. And like chopping firewood, the work will warm you twice.

The winter winds blow almost invariably from the north and west, and the north and west faces of your home will feel those icy blasts the most and will also receive less warmth from the sun. To protect foundation walls, earth berming on the north and west is especially effective.

First line the walls with polyethylene sheeting, which will create a moisture-resistant barrier to prevent the earth's dampness from being transmitted to the foundation wall. Then simply pile on the earth. To beautify your earth berm, exercise your green thumb. Plant grass seed or lay sod. You can even add to the insulating power by landscaping the berm with small evergreen bushes, which will further shield the house from wind and rain.

As a last resort, you can create the cheapest and quickest insulation for a foundation with a combination of fall leaves and polyethylene. Lay the plastic sheeting against the house, pile up leaves all around and cover the mound with more sheeting. Or fill large trash bags with leaves and distribute them around the perimeter of the house. It may not be the most attractive insulation, but it is effective.

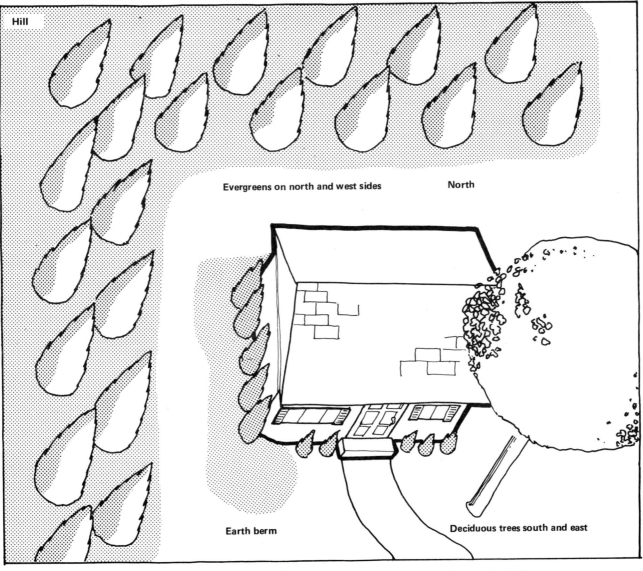

Hill

Evergreens on north and west sides

North

Earth berm

Deciduous trees south and east

A properly sited home

Let nature work for you in other ways, too. Selective tree planting makes a great difference, both as a shield against the winter wind and filter for the summer sun. On the north and east sides of the house, evergreens will create a year-round barrier to winds and storms. On the south and west sides, you should plant deciduous, or leaf-dropping, trees. During the winter these trees will let the sun shine through the bare branches onto your house for extra heat; during the summer, leaves will filter the sun and shade your house. Both sets of tree will deflect winds.

Windows

If we wanted to be absolutely sure that we'd be snug and warm all winter long (and cool throughout the summer), we would all live in climate-controlled airtight boxes. We would automatically save up to 70 percent on our heating bills if we lived in homes that had no windows and no doors.

But the human being, after all, is a sociable animal by nature and wants to know what's going on in the outside world. We use doors to let ourselves and our friends in and out, and windows to give us light, fresh air, and constant views of the world.

However, windows are also a house's biggest winter heat waster because glass is a remarkably fast conductor of heat, and it radiates the sun's heat back to the outside.

If you are building a new house, a careful window-placement plan will cut down on fuel costs. The sun can work for you as an energy source when maximum fenestration—the biggest expanse of glass—is on the south face. At the same time, cut back on windows on the other three sides of the house, especially the north face, but don't sacrifice light and the view.

While we are on the subject of siting a new house, try to position it so that it can be protected on its north side by a slight rise in the land or by a hill if possible, to act as a weather buffer. Then plant trees and berm the earth accordingly. If you study some of those old New England farmhouses, you'll notice that most of them are built on the south face of a hill, where they nestle against the earth protected from those north winds. Old evergreens also shelter those north faces, which are invariably shut tight—almost windowless.

One way to get more light and some extra heat through the window is to install skylights and clerestory windows (those that pierce the uppermost portion of the walls). Light will penetrate your rooms more effectively from a high vantage point; you will save on electric bills.

As you walk around your house site, select pieces of the view that can be enclosed by windows, but still lessen the overall fenestration wherever you can—unless you are planning a solar heating system. When you are thinking about the total window layout, consider fixed panes versus windows which open and close. If properly installed and sealed, fixed windows virtually eliminate drafts coming in or heat escaping to the outdoors. As much as 30 percent of your heat loss can occur through this infiltration—above and beyond the 30 to 40 percent lost through the glass itself. Air infiltrates a house through every crack and crevice; many of these exist around windows and doors, although you will also find them at any point where two surfaces meet, such as along baseboards and even around electrical outlets.

Infiltration is caused by the natural movement of wind around the house, the imbalances of indoor and outdoor temperatures and air pressures. Air-pressure fluctuations are caused by exhaust mechanisms such as chimney flues or appliance exhaust fans. In the process of

---

Broken Windows

1. If there is a small break in your window, you can block it temporarily with heavy paper, shirt cardboard, corrugated cardboard (from a broken box, perhaps). If possible, tape the material to both sides of the window; tape it strongly on every edge of the patch.
2. If the whole pane has been broken, cover the frame with a tightly stretched heavy blanket, tablecloth or other large piece of material. A sheet of 4 mil polyethylene can also be stretched across the window.
3. If you are somewhat handy and have the materials available, you can make a temporary "storm window." Make a frame the size of the window out of 1 ft-by-3 ft (30 cm by 90 cm) pine boards and staple polyethylene or another heavy plastic over it. This is par-

ticularly good for apartment dwellers, since it is relatively inexpensive and can be reused wherever you move.
4. If you are very handy and have the materials available, you can reglaze the window.
   a. Remove old pane by chipping away the putty and lifting out the broken glass. Wear heavy gloves to protect your hands.
   b. Clean the groove into which the new pane will fit.
   c. Measure the area of glass needed; it should be 1/8 in (.3 cm) narrower and shorter than the space into which it fits.
   d. Put a layer of putty all around the edges of the groove where the pane fits.
   e. Press the new pane into the putty.
   f. Hammer glazier's points (brads) at each corner to hold the glass in place.
5. A small crack or bullet hole in glass can be temporarily patched with duco cement.

Allow each application to dry thoroughly, adding layers until the hole is filled.

—E.L.

achieving equilibrium, air moves from areas of high density to ones of lower density. Permanently closed windows will minimize infiltration, but they can also breed feelings of claustrophobia. Most people want to be able to open windows, if only to get a breath of fresh air.

There are several types of operating sash windows, all of which will allow ventilation—and some infiltration. With any of them you can consider adding insulated shutters to the outside of the house. During the winter, they can be left closed over windows which aren't essential to your view or light, or they can be closed at night, when the room is unused or when the weather is at its worst.

Double-hung and horizontal sliding windows. One pane moves in front of the other; in a double hung, the lower pane moves up in front of the upper pane; in a horizontal sliding window, the left-hand pane generally moves in front of the right-hand pane.

Casement windows. These open out on hinges away from the window frame.

Awning and hopper windows. These are small ventilating windows which are usually installed below a fixed glass window. They hinge away from the frame, the awning toward the outside and the hopper inward.

Jalousie windows. These operate much like Venetian blinds. Many glass slats tilt outward by means of a crank. Jalousies are useful in warmer climates, since the slats can direct the air flow. They are impractical in colder climates because they might not seal tightly enough when closed.

All these windows are framed with either wood, aluminum, steel or a combination of metal and wood. Wood has a high insulating value, but if used alone, it must be sealed or treated with a preservative to repel moisture. Aluminum is a radical temperature conductor; it is corrosion resistant, and is best used in combination with wood or with an insulated backing. Steel is a poor insulator and corrodes easily, but it is very strong, and can be used in conjunction with an insulating material if properly treated to be weatherproof.

The choice of an operating sash window is really a matter of taste; awnings in conjunction with fixed glass will inhibit heat loss, but you will probably want some windows that could be used as emergency exits from your house, and another type of window with a bigger opening should be selected.

The major factor to consider when buying any type of window is its glazing. A single pane of glass will allow heat to escape very quickly; extra glazing stalls this process. Double- and triple-glazed windows, which are also called insulating windows, are now being recommended for most new buildings; their use may soon become law. Their insulating value stems not only from the extra glazing, but also from the sandwich of air between. Conversely, double and triple glazing, because of inhibiting conduction of heat through the layers of glass, keeps your house cooler in summer as well.

If you have an old house, the best investment and the greatest fuel savings can be realized by replacing all your old windows with new double- or triple-glazed windows (whether fixed or operating). A second choice is the addition of storm windows for extra insulation; but be sure to calculate savings versus investment before you decide on storm instead of new windows.

If you do opt for storm windows—or even if you already have triple-glazed windows—you will still need to give the regular windows a thorough checking to see if there are any leakage problems. No storm window, whether permanent or temporary, should be installed without

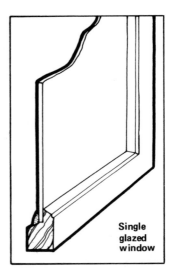

Single glazed window

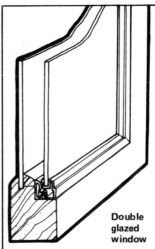

Double glazed window

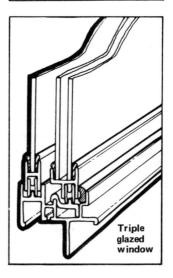

Triple glazed window

this check. Simply run your hand around all the edges of the windows, or move a candle around the edges. If you feel air on your hand, especially on a windy day, or if the candle flame flickers, there is infiltration. Even if you live in a rented apartment and are not directly concerned with fuel costs, you will want to keep that cold air out and the warm air in.

Weather stripping will create a proper seal on any window. It is made from a variety of materials, some of which are more durable than others, but all must have a flexible or resilient part which can be compressed to seal off the air flow. Before installing any weather stripping, make another quick check to be sure that the window is working properly. When it moves or swings open there should be as little friction as possible against the weather stripping. Make any necessary repairs, and then choose one of the following varieties of weather stripping:

Spring-metal. Virtually invisible when installed, spring-metal fits into side channels, across top and bottom rails of moveable frames and against the center cross strip (on a double-hung window). Spring-metal weather stripping comes with a nailing flange that has nailholes punched out at intervals for ease of installation. To tighten the seal on the spring-metal strips, slip a wide-bladed tool behind the unfastened side and bend it out slightly until the window holds tightly.

Tubular or foam-filled gaskets. Made of vinyl or rubber, these are mounted outside the window frame and will remain visible. These gaskets should be installed along every edge of the window frame, including the inside lower edge of the top pane of a double-hung window to seal the crack between both portions of the window. When installing, fit the thickened edge snugly against the window sash, and pull tightly as you nail across the strips. There is a special type of gasket for casement windows that has a deep groove to slip over the edge of the window frame. This weather stripping is installed with a special metal-to-vinyl adhesive.

Adhesive-backed felt or foam. Felt is the least durable weather stripping because it tears easily. It is installed in the same way as gaskets. Adhesive-backed foam is not very durable either because it can jar loose wherever there is friction, so it is best to install it only on the rails, which are friction free.

If you have taken all these precautions with your windows and still have infiltration, you probably have cracks and crevices around the outside of the window frames. These cracks can be filled with caulking compounds or sealants. You can choose compounds with an oil or resin base, a natural or synthetic rubber base or a purely synthetic base. The oil-based caulking compounds are the least durable and can stain. Of the synthetics, certain compounds—such as the latexes—must be used indoors because they deteriorate in the sunlight. All the other synthetics, such as butyl rubber, neoprene and silicone, are relatively easy to apply once you learn to handle the special caulking gun properly. They can work on any material in any location. Neoprene, however, is toxic and must be used where there is adequate ventilation. These sealants can also be applied with a caulking knife, and in some cases the compound is available in rope form and can be pressed in place with your fingers. But however you plan to apply a sealant, the surface to be sealed must be clean and dry, and the outside temperature must be above 10°C (50°F). Some sealants must even be warmed up before use, no matter what the outside temperature.

The caulking gun holds the sealant cartridges. Once the cartridge is set in the gun, cut the tip of the nozzle on the end of the car-

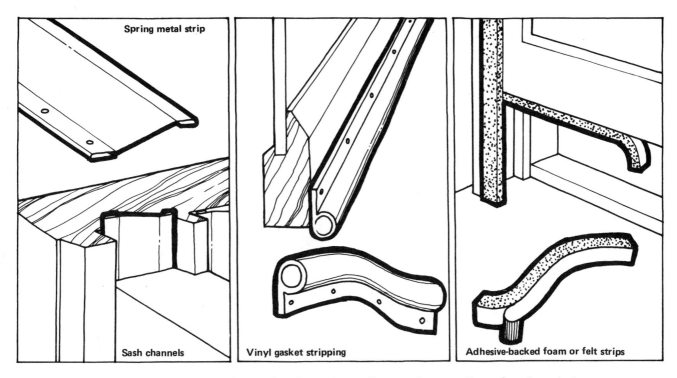

Spring metal strip

Sash channels

Vinyl gasket stripping

Adhesive-backed foam or felt strips

Types of weather stripping

tridge for a thin, medium or heavy flow of sealant, depending on the size of the cracks you are going to fill. The tip should be cut at a 45-degree angle for ease of application into the crack. A trigger on the gun releases the sealant; the pressure inside the cartridge keeps the sealant moving into the crack until you halt the flow by turning the plunger rod at the back of the gun so that its teeth point up, and then pull the rod back a bit.

If the crack you want to seal is very large, you must fill it to within a ½ in. (1.3 cm) of the edge with sponge rubber or oakum before squeezing in the sealant. For smaller caulking jobs, it is easier to use small roll-up tubes, which look like toothpaste tubes and which you control just by the pressure of your fingers. You can also seal small cracks or areas with the rope caulk; use glazing compound for the cracks in the window panes themselves.

You should also check the doors in the same manner. A few other areas are candidates for caulking: the corners formed by exterior siding; the junction of a porch and the house; where masonry or a chimney meet siding; where pipes or ducts penetrate siding.

Storm Windows. These have a pocket of air between layers of glass much like insulating windows, and can save you about 25 percent on your heating bills.

Storm windows are best installed by a professional, because they must fit snugly to work at all and many windows on older houses, even if they seem to measure to average sizes, will often have individual quirks that could cause air leakage.

All storm windows, however, should have a couple of tiny "weepholes" at the bottom to drain any moisture buildup after rain or snow. Weepholes also prevent condensation at the sill, which could cause rotting.

There are two types of "stopgap" storm windows which you can make and install yourself either inside or outside the house. For both types, you will need staples or nails and enough heavy-duty polyethylene sheeting to cover the surface of all windows.

The simplest do-it-yourself storm window consists of the sheet-

Older homes often have many cracks around doors and windows. Masonry exteriors—stone, stucco and brick—often develop cracks which should be sealed. But old or new, frame or masonry, insulated or not, all homes suffer some heat loss through convection. Remedies for this situation—caulking, weather stripping, and sealing—are among the least expensive of energy savers and, in most cases, are do-it-yourself jobs.

On the outside of the home, look for cracks in the following places:
a. around window and door frames, sills and joints (check the putty around window panes and repair if cracked or worn away)
b. at corners formed by siding
c. between porches and the main body of the dwelling
d. around water faucets, electrical outlets and gas or oil lines
e. where the chimney or masonry meets the siding

On the inside of the home, check:
a. between foundation and the sill plate
b. around ceiling fixtures
c. around water pipes and drains
d. around the furnace flue, plumbing vents, pipes, and air ducts in the attic
e. between heated and unheated areas, such as attached garages and crawl spaces

Caulking compounds are used to seal cracks and are available in cartridges to be used with a caulking gun and in bulk, tube and rope form. Choose the caulking compound best suited for your job.

Oil or resin base is the least expensive type of caulking compound. It is also the least durable, lasting as little as one year. Although it bonds to most surfaces, a primer must be used if it is applied to a porous surface. Oil or resin-base compounds cannot be applied successfully if the temperature is below 16°C (60°F).

Latex requires a primer if applied to a metal surface. It will last up to ten years if applied when the temperature is above 4°C (40°F). Latex is an excellent choice for wood and painted surfaces.

Silicone requires a primer if applied over a bonding agent or over a previously glued crack. It will last up to twenty years and can be applied in below-freezing temperatures. Silicone is excellent for use on wood, metal, and painted surfaces; its adhesion to masonry is slightly less effective.

Butyl requires no primer and is very durable, lasting up to twenty years. A butyl compound must be applied when the temperature is above 4°C (40°F). It is an excellent choice for unpainted wood, metal, and masonry surfaces.

Do not use lead-base caulking compounds; they are toxic.

With a little practice, the application of caulking material is fairly easy; but because a thorough job takes time, spending a little extra money for a durable compound is worth your while.

If you have any doubts, consult a reputable hardware or building materials supplier about the type of caulking compound required for your particular job.

To apply caulking compound, first thoroughly clean the area to be caulked with a solvent and putty knife. Get rid of dirt, built-up paint and cracked putty or old caulking compound. If cracks are deep or wide, fill them with caulking cotton, sponge rubber or mineral wool before applying caulking compound.

A caulking gun is useful for long, narrow cracks. Hold the gun at a 45° angle and work from top to bottom. Make sure the bead covers both sides of the crack. It may be necessary to apply a second bead if the crack is wide. Seal the two beads.

A warning: Don't caulk windows or doors so that they cannot be opened in case of emergency.

—From *Save Energy Save Dollars* (see bibliography)

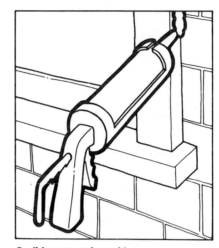

Caulking gun and cartridge

ing, alone. Cut the polyethylene to fit the window, allowing a hem of 3 in. (7.5 cm) all around. Roll the hem around itself several times at each edge and then nail or staple through the roll to the window frame so that the rolls make a good seal all around the window. Because this type of storm window is not very pretty to look at in your living room, you may wish to install them on the outside of the house. Hard-to-reach windows, however, will have to be protected on the inside.

You can also make plastic windows by first constructing simple frames which will fit against or just within your regular windows. Use furring strips and angle irons; measure for each window separately because of slight differences in size. Then nail or staple the plastic sheeting to the frame after cutting and rolling as described above. Screw the storm window to the window. In the spring, remove and store these frames. Carefully mark each one and the window it covered for easy reference the following autumn or winter.

Doors

Like windows, doors are heat wasters and must be effectively sealed with weather stripping. Although they are not commonly made of glass, they do lose heat through their base material. They also lose a great deal of heat every time they are opened and closed—and if they do not shut tightly, they will lose even more. If you can, close up and do not use as many doors as possible during the winter.

The main door to the house should be constructed of a heavy-duty wood or insulating material, and covered in another insulating sheath or weather-resistant material. Storm doors may be installed outside to set up another insulation barrier against cold and wind. When closed, the storm door keeps a layer of air sandwiched between it and the main door; when opened, it begins to close as the primary door opens. This alternating sequence of opening and closing blocks

---

Keeping Warm in Your Vacation House

1. The site and location of your house helps determine how warm you will be in the house. Strong winds can make a house very cold (especially one that is poorly insulated or that has a lot of windows on the windy side). In general, mountain winds blow up-canyon during the day and down-canyon during the night; sea winds blow away from the sea during the day, toward the sea during the night. Remember, the cool breeze in the summer may become an unpleasant chilling factor in colder weather. Cold air tends to settle in pockets or any depression in the terrain. Valleys or mountain meadows that are cool during the day may become unbearably cold at night. The site should be away from a wind gap—a place where the wind rushes through. A wind gap can be recognized by the fact that the trees and/or vegetation lean sideways, or if there is a lack of vegetation in an otherwise verdant area.

2. If you are building your second home or buying one that was recently built, the floor will most likely be either concrete or wood. Wood floors tend to be a bit warmer than concrete because wood is a less effective heat conductor. Building paper between the subflooring and the finished flooring (in a wooden structure) will serve as insulation.

3. Bottled gas (used in heating) is heavier than air. If gas leaks in a basement it will collect at the lowest point, and there is a risk of explosion. The safety valve on the bottles should be placed so that if the pilot light goes out, the gas to the pilot automatically will be shut off.

4. Conventional oil and gas furnaces need electricity to function. If your house is too far away to be hooked into the local utility company lines, consider installing your own gasoline- or bottled-gas-powered generator. This would be impractical for electric heating, but it is certainly a feasible way to satisfy other electrical needs (oil or gas burners, lighting, hot water). The least expensive type of generator is started like a power mower; more expensive models have a self-starter, similar to the self-starter in an automobile. There is a very expensive generator that will turn on as soon as there is a demand for electricity.

Wind may also be used to generate power for electricity, but this is a very iffy proposition. Because wind is intermittent, the power would have to be housed in storage batteries, and usually only a couple of hours can be thus stored. A standby gasoline-powered generator would be a necessity.

some of the leakage of heat and air.

The best air lock, however, is a little entry porch, for it increases the distance between the two doors and thereby insures that one door will be closed while the other is open. You can easily construct a small entry porch by building a frame from two-by-fours, slightly wider and higher than the outside or storm door, allowing about a 4 ft. (1.2 m) space between storm and primary doors. Add roof trusses, with both ends angled to fit each other and the frame. Sheathe the whole structure with ¾-in. (1.8 cm) plywood and then cover with the siding and roofing of your choice. The little porch need not be insulated.

The junction between the air-lock entry and house should be sealed and, of course, the frames around both doors should be checked to see if they need any caulking.

The main doors may need weather stripping too. Seals for the sides and tops of the doors are different from those used along the threshold. However, before you embark on weather stripping, check the door to see that it is functioning properly. If it is scraping too much at the bottom or top, you may have to remove the door and reset the hinges.

Most of the air that escapes around the door leaks out under the bottom, so this is the first area that you should protect. To block the bottom crack, use weather stripping sweeps or threshold strips.

Sweeps, either plain or spring-operated, pull a flexible material against the threshold. They can be installed to the outside bottom edge of the door without removing it from its hinges.

The threshold strip is a bit more complicated. It consists of a flexible material, such as a vinyl bulb, which is attached directly to the threshold and presses upward against the bottom of the door. They are more difficult to install snugly and correctly, and the door must be removed from its hinges. Some strips are interlocking, with one section attached to the threshold and the other to the bottom of the door.

The V-strip is doubled-over metal and is extremely durable. It is installed between the door frame and jamb. The nailing flange, with nailholes punched through at intervals, fits against the doorstop and is nailed into the jamb.

Odds-and-ends. Plastic tubing, foam-edged wood or metal or plastic strips and adhesive-backed foam strips are all affixed directly to the door frame or doorstop molding. The flexible edge of any of these materials presses against the door, creating the weathertight seal.

Now, with the door firmly shut against the cold winter winds, the windows sealed tight and the house wrapped in a protective coat of insulation, your home can settle down to its long winter's nap and you are free to enjoy the elements when you want to, and not as uninvited guests into your home.

Closing Up Your Summer House for the Winter

If you are among the millions of second-home owners who spend only summers and warm-weather vacations in your getaway house, you probably prefer not investing in insulation to make the house habitable for winter living.

In this case, shutting the house up for those cold months is as much of a ritual as getting it ready for summer and is more important to do carefully to avoid a self-igniting fire or burst pipes.

The most important freeze-prevention maneuver is draining out all the water pipes and preparing them to withstand subfreezing temperatures without bursting. Before you get ready to drain the system, thoroughly flush it out to get the pipes as clean as possible. Use a chemical cleaner, either liquid or granular, in all sinks. When the bubbling of decomposing vegetables, dirt and accumulated goo has subsided, flush lots of water through the pipes. Then pour one box of baking soda down each sink and flush that through. Baking soda refreshes the pipes and rids them of any greasy residue.

Once the pipes are clean, drain them. Place large buckets under elbows of piping at their lowest point, at drain valves of the hot and cold water lines and also under valves on the storage tank or water-heater source. When all the buckets are in place, open all the faucets in the house, and then open all the valves. Drain the water out of the pump, too; there is a plug at the base.

Once you have emptied the pipes, you will have to clear the storage tanks of the toilets. After everything is empty, pour kerosene or antifreeze into every sink and toilet. The kerosene will mix with the water that is left in any traps to resist freezing temperatures and prevent water pipes bursting at these points. You will need about 1 quart (1 liter) of kerosene for the sinks and shower, 2 qt. (2 l) for each toilet.

Leave all the faucets open for the duration of the winter.

If you have a heating system in the summer house, you can shut it down at the main switch, which is usually clearly indicated at the furnace. It is best to have the fuel supplier check your system at this time; he or she can shut it down for you if you like, and start it up again in the spring or summer.

If you have a canister-gas setup, arrange to have the last canisters taken away by your supplier before you leave for the season.

Furniture and linens fall victim to mildew during the cold months. Pillows and mattresses can be propped up slightly for some ventilation. Blankets, bedspreads and other bedcoverings can be left on the bed; they will stay fresh if they are thoroughly cleaned. If you decide to keep them on the bed, drape them tent-style over the propped-up mattress so that they will get some air.

Pull upholstered furniture away from windows so that the sunlight will not fade the fabrics while you are away. Better still, if you have draperies at the window, pull them closed most of the way to prevent too much exposure to sunlight.

Rugs should be cleaned; if they are cleaned professionally, they can be wrapped tightly to prevent mildew. Otherwise they should stay flat on the floor, away from sun.

If you have sheets that you use for summer and don't care about too much, you can cover your furniture with them to keep off sunlight and dust.

Sheets should get air through the winter; you can leave them stacked in the linen closet with the door slightly ajar. If you think they may get dusty, you will have to seal them tightly; just closing the door on them will not afford adequate protection. Tie them up in plastic bags, making sure no air at all will infiltrate.

Food will attract rodents and insects. Anything in boxes should be eaten up or thrown away. Canned goods can stay, because they are vacuum sealed and can withstand freezing temperatures. Likewise, liquids that don't freeze, such as oils and liquors, can stay without harm. Other bottled goods should be disposed of, as they may burst.

Clean all appliances to thwart insect infestation. Leave the refrigerator slightly open; it and all other electric appliances should be disconnected.

If you have a fireplace, make sure the damper is closed and the fireplace chimney is clean. Check the chimney cap to see that it will prevent rain and snow from entering the chimney.

Windows and doors should be tightly shut. If you have large expanses of glass, especially near the ocean or other windy areas, put up plywood sheets or shutters to protect the glass during severe storms. All windows are best protected with shutters.

Any gutters or downspouts should be cleaned before leaf-dropping season. Strainers will catch leaves and prevent them from clogging the downspouts.

Alert the post office and local fire and police departments when you leave; they will usually keep tabs on your house in your winter absence and will report any damage due to weather or vandalism.

More tips on winterizing:
1. Make sure that everything left outside is secure against high winds, sleet and snow and ice.
2. Take all screens off doors and windows and store them inside. Grease all hinges, levers and openers to keep them from rusting during the winter (especially if your home is near the sea.) Lock windows from the inside if possible, and nail stripping around the doors (top and bottom) if there might be heavy winds that will blow in sand and dirt. Sheets of plastic may be tacked over windows and doors.
3. Leave dresser drawers slightly open; air circulation will prevent mildew.
4. After removing rugs and carpets from floors, clean the floors and wax and polish them—a coat of wax will prevent damage from moisture and cold weather.

Heating Systems

Gone are the days when the major source of heat in the home was the fireplace or woodstove, and when the family huddled together before the hearth on cold winter nights or warmed their hands over a few lumps of coal. The home heating system has become increasingly efficient and elaborate, but it is still composed of two major elements: an energy source to create heat and an apparatus to circulate the heat through the house.

Energy Sources

Some 310 million years ago, in the Paleozoic era, fossil fuels were formed in the earth's layers; but it wasn't until a mere 3,000 years ago that these fuels began to be used. The Chinese first burned coal to smelt copper.

Today, three major forms of fossil fuel—oil, natural gas and coal—are our chief sources of energy. After World War II, oil became number one, and fuel became something we took for granted. It was plentiful and it was cheap. No longer did we have to suffer through the inconveniences of coal, which had to be painstakingly mined from the earth and burned in furnaces which needed daily cleaning. Oil—aptly nicknamed "black gold"—flowed almost effortlessly from deep within the earth, was transported quickly and easily to refineries and from there to our homes, where it powered the furnaces and provided heat. Then came the oil embargo of 1973 and the subsequent price hikes, which jarred us into realizing some hard facts: fossil fuels are finite; the United States has grown increasingly dependent upon foreign sources for fuel; we have been wasting our own fuel resources.

Even so, the fossil fuels still generate 95 percent of our energy and any discussion of fuels must begin with them. The following list will give you thumbnail profiles of our main energy sources, beginning with the fossil fuels. Each fuel is rated for its thermal efficiency. No fuel is 100 percent efficient. Fuels are combustible; the process of burning releases gases and fumes that must be vented. Therefore, some of the heat generated by the fuel is inevitably lost out the chimney or flue attached to the home heating plant. The thermal-efficiency index gives the percentage of usable heat produced by the fuel under good conditions. And to be most efficient as well as safe, fuels must be burned cleanly. This means that they should be burned in sufficient oxygen and in properly maintained equipment, so that the flames are always at the highest possible temperature and there is the least amount of smoke. Smoke is actually the by-product of improper combustion, and is composed of unburned bits of carbon and droplets of tar.

The fossil fuels:

Oil: Until the embargo, oil was plentiful and inexpensive. It is easy to transport and to store, which makes it attractive as a fuel source. Oil has a thermal efficiency of between 40 and 60 percent. It must be burned properly in a well-maintained furnace or it will give off pollutants. Sold by the liter (gallon).

Gas: This fuel comes in two forms—natural and liquid gas. Natural gas is difficult to transport and store. The imported and domestic supply, like oil, is diminishing. Natural gas burns clean and is not noxious. It has a thermal efficiency of between 50 and 70 percent, and is sold by the measure of 1,000 cubic feet (mcf).

Liquid gas—butane or propane—has the same thermal efficiency as natural gas. It is sold in canisters, and is especially convenient as a fuel source for out-of-the-way places, mobile homes or vacation homes.

Coal: Although bulky and dirty to transport, store and burn, coal is regaining popularity as a fuel source, primarily for large-scale industrial operations rather than for domestic use. The reason for this resurgence is that coal is the one fossil fuel that is still plentiful in this country. It is, however, hazardous to extract from mines deep in the earth, and the alternative—strip mining—destroys the landscape. Coal has a thermal efficiency of about 55 to 65 percent. It must be cleanly burned or it will give off lethal carbon monoxide and other pollutants. The ashes must be removed from the furnace daily.

Other energy sources include:

Electricity: Electricity is a prime energy source, not a fuel. It is easily transmitted; it provides heat as it travels over high-resistance wires or coils. The friction it sets up stimulates heat. Electricity has a thermal efficiency of about 35 percent and is sold by the kilowatt or kilowatt hour (KWH). It can be used in combination with other energy sources, such as solar power.

Solar energy: Of all the alternative energy sources currently being researched and developed, solar energy is probably the most feasible. It will make a substantial difference in heating and will provide energy savings. At the present time, solar energy is most commonly used for heating hot water in the home, but it can be used for heating space too. Solar heating systems may be "passive" or "active." Passive systems use the sun's heat naturally with little mechanical hardware, expense or power, while active systems are more elaborate, with a more mechanized distribution system to get heat into the house and store it. Solar research is also focusing on silicone cells, which can concentrate the sun's rays many times over, and also can convert solar energy directly into electricity.

Wood: This was one of the first sources of heat; until 1900, almost half the energy consumption in the United States was from wood. With the emergence of coal, followed by oil and gas, wood was rapidly supplanted as a primary fuel source. However, the energy crisis has brought wood back into favor, to supply fireplaces, wood stoves, and occasionally even wood furnaces. Wood is sold by the log or by the cord (a pile of logs which measures 4 ft. by 4 ft. by 8 ft.).

Wood is a renewable resource and the lumber companies are involved in careful regeneration of their timberlands to maintain a correct balance between areas that are cut and areas that are cultivated. They are also actively involved in research to develop fast-growing trees and to create clones. The National Forest Service and Bureau of

Land Management encourage limited cutting of fuel wood in the national parks. Permits are issued for specified areas allowing 10 cords per person per season for home use. The only stipulation is that you take only cut-down trees—also called "cull" trees—which are twisted and unsuitable for building and other commercial purposes. Taking away these trees leaves room for new, healthy trees to grow in their place.

Ten cords is actually about double the amount the average total-wood-burning six-room house would need each winter. And, of course, most houses use wood in addition to another energy source.

The best woods for burning are the high-density, heavier woods, such as oak and locust, but any good wood will burn, unless it is totally sodden. Wood should be aged, or allowed to dry, for about a year, to burn most efficiently. A good dry log, especially one of high density, will give over 40 percent more heat than a damp green log.

Heating Systems

Almost one-quarter of our total energy consumption goes into heating (and cooling) our homes. Many houses are heated by forced hot air or forced hot water or steam: systems fueled by oil or gas. Electrical heating systems, which save space because they need no chimneys, are gaining by about 7 percent a year.

Whatever your heating system, here are several general words of advice. To increase the warming power of your heating system, add registers (openings in the floors or walls) to set up a convection process. This makes use of the natural rising and falling of heated and cooled air to circulate the air from lower to upper floors and back. These added registers should be separate from those attached to heating ducts. Place registers near the inner walls of the room to bring the hot air upward, place registers through which cold air will descend

---

Firewood Ratings

The table below shows the relative ratings of a variety of dried woods.

	Relative amount of heat	Easy to burn?	Easy to split?	Does it have heavy smoke?	Does it pop or throw sparks?	General rating and remarks
Hardwood Trees						
Ash, red oak, white oak, beech, birch, hickory, hard maple, pecan dogwood	High	Yes	Yes	No	No	Excellent
Soft maple, cherry, walnut	Medium	Yes	Yes	No	No	Good
Elm, sycamore, gum	Medium	Medium	No	Medium	No	Fair—contains too much water when green
Aspen, basswood, cottonwood, yellow-poplar	Low	Yes	Yes	Medium	No	Fair—but good for kindling
Softwood Trees						
Southern yellow pine, Douglas fir	High	Yes	Yes	Yes	No	Good but smoky
Cypress, redwood	Medium	Medium	Yes	Medium	No	Fair
White-cedar, western redcedar, eastern redcedar	Medium	Yes	Yes	Medium	Yes	Good—excellent for kindling
Eastern white pine, western white pine, sugar pine, ponderosa pine, true firs	Low	Medium	Yes	Medium	No	Fair—good kindling
Tamarack, larch	Medium	Yes	Yes	Medium	Yes	Fair
Spruce	Low	Yes	Yes	Medium	Yes	Poor—but good for kindling

(Courtesy of Preway.)

near the outer walls. Hot-air registers can be placed near sources of heat as well, such as the furnace, fireplace or wood stove.

If your heating system includes a furnace, you can reduce your fuel bill by as much as 15 percent through proper maintenance. Servicing may be written into the contract with your fuel supplier, but if not, make certain that your equipment is inspected, cleaned and repaired once a year, preferably before the cold weather begins.

The service person will conduct a number of tests to be sure the furnace is operating at maximum efficiency. For an oil burner this means four things: a draft test to check for excessive heat escaping up the chimney and for suffcient draft to burn the oil; a smoke test, to check for clean burning; a carbon dioxide test, to be sure the oil is burning completely; and a stack-temperature test, to check the temperature of gases moving up the stack. With gas furnaces, the main gas and control valves are checked, as is the pressure regulator.

A handful of heating systems:

Forced-hot-air heating system. This consists of a burner or furnace that heats the air and ducts through which the hot air rises into

A forced hot air heating system

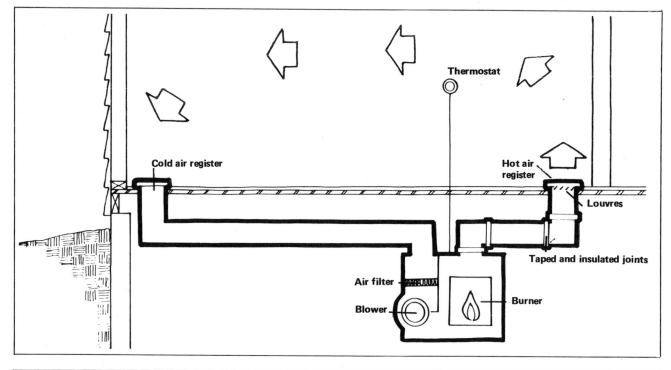

Keeping Warm in Your Apartment

If you rent an apartment or any other space in a multiple-unit dwelling and you feel your landlord is giving you less than adequate heat or hot water, here are some things you can do.

Most cities and counties have ordinances that specify the minimum amounts of heat and hot water a landlord must furnish tenants. The agency that enforces these ordinances usually contains the word *housing, building* or sometimes *health* or *environmental health.* If you do not live in one of the cities on the chart below, try to find these key words in your local yellow pages under "Government." When you get through to the right number, chances are your complaint will be taken seriously—one of the main jobs of building inspectors is to

enforce the ordinances—but you will get faster results if you can supply temperature figures. If results are not as fast as you would like, remember that the cold affecting you is affecting a lot of other people in the same area (on a record day, over 3,400 people complained to New York City inspectors).

Although they differ, most ordinances provide for a minimum indoor temperature during the day and a lower one at night, during certain months of the year or whenever the temperature outdoors falls below a specified temperature. Indoor temperatures are measured in the center of the room at a height of 3 ft. (90 cm) from the floor. Inspectors use thermometers coupled with clocks that give a reading over several hours.

Hot water is not always under the jurisdiction of the building inspectors. If it isn't, it is probably a problem for the health department. Most ordinances specify that the landlord shall provide hot water in every

kitchen sink, lavatory basin and bathtub. Hot water is generally defined as being at least 49°C (120°C).

See the chart on page 35 for the ordinances of major cities in the Northern United States.

If the place where you live seems not to be covered by anyone's building-maintenance ordinance and you are sure that the indoor temperature in your rented space is below the temperatures on the chart, call the health department. There is a good chance that even without an ordinance they will be able to tackle your problem as a potentially hazardous health condition.

CITY	HEAT				HOT WATER	AGENCY AND PHONE NUMBER	NOTES
	Day	Night	When Outside Temp. Is Below	Months			
Anchorage	20°C 68°F	20°C 68°F		At all times	No specified minimum	Department of Environmental Protection 264-4720	The department spokesperson defined hot water as too hot for the inspector to keep his hand in
Boston	20°C 68°F	18°C 64°F		9/15 to 6/15	49°C 120°F	Housing Inspection 725-4785	Day: 7 A.M. to 11 P.M. Night: 11 P.M. to 7 A.M.
Chicago	See note	13°C 55°F		At all times	49°C 120°F	City Building Department Complaints 744-3420	15.5°C (60°F) at 6:30 A.M. 18°C (65°F) at 7:30 A.M. 20°C (68°F) from 8:30 A.M. to 10:30 P.M.
Cleveland	21°C 70°F	21°C 70°F	10°C 50°F		49°C 120°F	City Hall 694-2825	
Denver	See Note	See note	See note	See note	49°C 120°F	Environmental Health Services Housing Section 893-6144	No ordinance covering heat supply, but system must be capable of supplying 21°C (70°F) when outdoor temperature is -20°C (-5°F)
Detroit	21°C 70°F	18°C 65°F	13°C 55°F		49°C 120°F	Housing Division 224-3105	Day: 7 A.M. to Midnight Night: Midnight to 7 A.M.
Indianapolis	20°C 68°F	20°C 68°F	13°C 55°F		49°C 120°F	Bureau of Environmental Health, Housing Section 633-3644	
Milwaukee	19°C 67°F	19°C 67°F	See note	See note	No minimum	Building Maintenance	No minimum for hot water. When outside temperature is below -12°C (10°F), minimum requirements are relaxed
Minneapolis	21°C 70°F	21°C 70°F	15.5°C 60°F		49°C 120°F	Housing Complaint line 348-7858	
New York City	20°C 68°F	13°C 55°F	13°C 55°F		49°C 120°F	Central Complaint 960-4800	Day minimum temperature applies when outside temperature is below 13°C (55°F). Night minimum temperature applies when outdoor temperature is below 4.5°C (40°F). Day: 6 A.M. to 10 P.M. Night: 10 P.M. to 6 A.M.
Omaha	See note	See note	See note	See note	49°C 120°F	Health Nuisances 444-7481	Landlord must prove capability to heat building to minimum 21°C (70°F) under minimum winter conditions, but no ordinance requires delivery of heat.
Philadelphia	20°C 68°F	20°C 68°F	See note	See note	49°C 120°F	Department of Licensing and Inspection Abatement Unit 686-2588	Minimum temperature applies from October 1 through April 30, and in May and September when outdoor temperature is below 15.5°C (60°F)
Portland	20°C 68°F	20°C 68°F	20°C 68°F		49°C 120°F	Health Department 248-3400	Hot-water ordinance covers capability only, but health department would consider temperature below that a violation.
Providence	19°C 67°F	15.5°C 60°F		10/1 to 5/1	49°C 120°F	City Hall Building Inspector 421-7740	Day: 6:30 A.M. to 11 P.M. Night: 11 P.M. to 6:30 A.M.
Seattle	18°C 65°F	14.5°C 58°F		10/1 to 5/1	49°C 120°F	Police Department Community Service Program 625-4661	Day: 7 A.M. to 10 P.M. Night: 10 P.M. to 7 A.M.
Washington, D.C.	20°C 68°F	18°C 65°F		At all times	49°C 120°F	Complaint Center 724-4414	Day: 6 A.M. to 11 P.M. Night: 11 P.M. to 6 A.M.

the house through registers in the walls or floors. You can control the access of heat into every room by manipulating the louvers on the registers. A hot-air system works best when the ducts are properly insulated and taped at their joints, and the air filters are kept clean. When the warmth in the house reaches the level indicated on the thermostat, the hot air shuts off immediately. The system starts up again when the temperature drops below that level. A more efficient hot-air system circulates the air continuously. You can convert an on-off system to CAC (continuous air circulation) by replacing the motor with a special two-speed motor and control. Once a year, check the hot-air system and clean the filters and blower.

Forced-hot-water heating system. This consists of a burner or furnace, boiler, pipes and radiators. It is a more elaborate system and provides more even heat than the hot-air system, since water retains and radiates heat longer than air. The furnace heats the water in the boiler and circulates it through the pipes and radiators. Cold water returns to the boiler through another set of pipes. Forced-hot-water systems should be routinely checked to be sure pump, valves and motor are running smoothly. Once a year "bleed" the radiators to let accumulated air escape. You should also drain the boiler to prevent sediment buildup.

Forced-steam heating system. Almost exactly like the hot-water heating system, except here the water in the boiler is changed to steam before it is circulated to the radiators. Each radiator has a steam valve

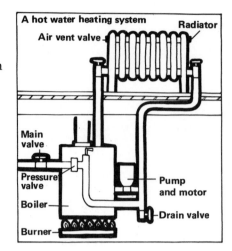

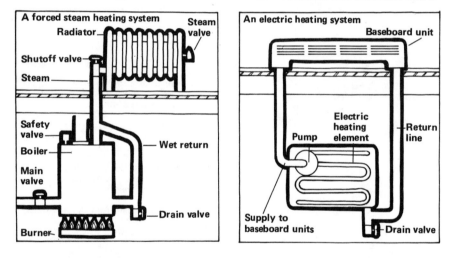

on it. Look after the system just as you would a hot-water heating system.

Electric boiler. Also called a hydronic system, this method needs no chimney. It transfers heat from electric immersion elements to water, which is then circulated through pipes to baseboard units. Some systems operate at off-peak hours, when the electrical rates are lower, and store the hot water to be used during the day, when electrical rates are higher. An electric boiler makes a convenient replacement for a conventionally fired boiler. It takes up only about 1 cu. m. (3 cu. ft.) of space, but operates with the same distribution network as the steam heating system.

Electric furnace. This consists of a series of heating elements that function in sequence to create a buildup of heat to desired temperature. A blower fan moves the air through ducts much like those in the conventional hot-air system. The electric furnace can also be coupled with a central air conditioning unit to make year-round use of the same ductwork throughout the house. It needs no chimney. An electric furnace can be installed anywhere convenient, either on the floor or ceiling of the basement, because it is both compact and fairly lightweight.

Wood furnace. This is not as commonly used as the preceding systems, but it is coming back, especially as a supplementary unit. Similar to the circulator fireplace (see pages 41-47), the wood furnace operates on a much larger scale by burning woods in much greater quantities. Some wood furnaces come with auxiliary burners, using either oil or gas, which can be switched on as a back-up system when wood is in short supply or which turn on automatically when the wood runs out altogether. A wood furnace heats the air circulating around the firebox; the air is then forced through air ducts into rooms of the house. Some types circulate water in much the same way, distributing it through pipes into radiators.

Space heaters. Not a total heating system, the space heater is a

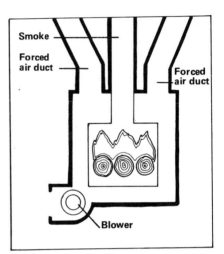

A wood furnace

No Water—What to Do about Frozen Pipes

1. Open several faucets in the house before trying to thaw frozen pipes. As the pipe begins to thaw, steam and water will begin to build up; if they cannot escape through an open faucet, they may cause the pipe to burst.
2. Start heating pipe close to an open faucet.
3. Heat can be applied with a heat lamp, a photographer's lamp, a hair dryer, an iron or a vacuum cleaner (with the hose attached to the exhaust hole). If possible, rig a stand for the heat source; it may take hours to thaw a frozen pipe.
4. The quickest way to thaw a pipe is with a propane torch. Keep the torch moving and remove it from the pipe every now and then to prevent the buildup of steam in one location. An asbestos shield should be used to protect the wall behind the pipe from the flame. Burning newspaper can be used as an emergency "torch."
5. A slow, messy way to thaw pipes is to wrap them in hot, wet towels or rags; saturate the cloths continually with boiling water. Boiling water can be poured directly on the pipes. In either case, have a basin underneath to collect the water. This method, of course, assumes the availability of an alternate water source.
6. When thawing a drain pipe with direct heat, start at the lower end, so that the

melted water will flow off. Chemical drain cleaners can also be used to thaw drains, and sometimes just the constant flow of hot water will do the trick. Pouring salt down the drain first may speed the process.

7. If the pipe is underground or otherwise inaccessible, more tedious work will be required. The frozen pipe should be opened on the house end; insert a small pipe or tube. Pour boiling water through a funnel into the small pipe and keep pushing the pipe through as the ice melts. To ensure a good flow through the funnel, hold it higher than the frozen pipe. If necessary, keep on adding extra pipe or tubing to the thaw pipe until there is a clear flow of water through the formerly frozen pipe. If the frozen pipe is extremely long, a small force pump will accomplish the job much more efficiently than the funnel.

—E.L.

Alternate Water Sources—What to Do until Pipes Thaw

1. Water may be stored in large plastic bottles, filled at the beginning of winter for just such emergencies.
2. Snow and ice can be melted for water. A considerable amount of snow is needed to get one glass of water; however, ice, a more concentrated form of water, yields more liquid. Remember, if you plan to use the water for drinking, that ice is only as pure as the water that made it; freezing, unlike heat, does not kill germs. To make sure that your water is pure, boil it before drinking. Whiskey, by the way, though it might make your water tastier, will *not* purify it.
3. Urban dwellers without ready access to newly fallen snow might keep extra ice cubes in their refrigerators to be used as an emergency water source.
4. For drinking and cooking, your larder shelf may come in handy. Fruit juices and soups can be used as liquid in some cooking processes.

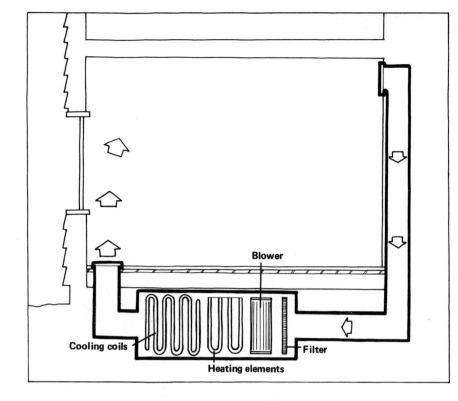

Blower

Cooling coils

Heating elements

Filter

portable device that is useful to supplement the basic heating plant. Its effect is limited because it usually heats only the concentrated area where it sits, but it will take the chill off an otherwise underheated space, and are particularly handy in workshops or garages that have no other heat source.

With increased regulations by the Consumer Protection Agency and other watchdog groups, gas- and oil-fired space heaters are being phased out of production by many manufacturers because they are dangerous and carry the inherent risk of explosion if improperly lighted. Because oil and gas emit fumes, space heaters that use these fuels must be vented to the outdoors.

Your best choice is the electric space heater. It requires no venting and is portable. It operates by convection: air passes over the heating element, rises, and is then deflected into the room via a curved reflective surface behind the heating element. Electric space heaters can get extremely hot and should be placed out of the reach of children.

Power Failures—A Checklist and Some Remedies

1. Check the fuses or circuit breakers. If you are using extra electrical appliances (lights, portable electric heaters, etc.) you may simply have blown a fuse Switch off all appliances and the main fuse before attempting to change a fuse. The blown fuse will probably have a black or brown spot on it and can be identified this way.

2. If the whole neighborhood is dark and there is reason to believe that the power lines are down, inform the utility company. Turn off all your electrical appliances to lessen the possibility of a secondary blackout when the power returns.

3. Your heating system is likely to be affected by a power blackout; check the "No Heat" chart in this section and the rest of the book to find ways to keep warm.

4. If the power outage lasts only a short time, the food in your freezer and refrigerator should keep. Try to open the door as little as possible so that the unit can maintain its temperature. Foods that must be kept cold can be placed outside if there is reason to believe that you are in for a long period without electricity.

—E.L.

As with all electrical appliances, these units should not be used near any water source—especially a bathtub.

When buying a space heater, check for a UL (Underwriters Laboratories) listing, see that its covering grill cannot be penetrated, and that the body and handle of the unit are insulated so you will not get burned if the heater is to be moved while it is still hot.

Heat pumps. The heat pump, like the space heater, is a supplementary heating device. Though it can work alone, it is most often used in conjunction with an existing energy source (especially electricity). In its forty years on the market, the heat pump has been thought of as a reverse-cycle air conditioner, because it operates on the same principle. In fact, some early versions were remodeled air conditioners, but they did not work very well. Today's heat pumps are much more efficient and can save about 25 percent of fuel costs. A heat pump gives off air warmed to about 40°C (104°F) which is some 15°C (27°F) lower than what many people are used to with their fossil-fuel-fired systems, but by no means uncomfortable.

The heat pump creates heat by depending upon the change of state of liquid to gas, and vice versa, within its refrigerant and coils. When gas changes to liquid under pressure, it gives off heat.

In a "split system," the heat pump consists of two units, one placed outside and the other indoors. It has five basic components: indoor coil, outdoor coil, reversing control or valve, compressor and refrigerant, plus fans and blowers to circulate the heated or cooled air.

When the heat pump functions as a cooling device it acts like an air conditioner, absorbing heat into the refrigerant in the indoor coil (because the refrigerant is colder than the indoor air), which then evaporates into a gas. This gas is drawn into the compressor, where it changes under pressure back into a liquid, and moves to the outdoor coil, where it condenses still more as it transfers the heat buildup to the outside air.

In the heating cycle, this process is reversed, so that the outdoor coil functions as the evaporator and the indoor coil as the condenser. Until recently, heat pumps could not function in below-freezing temperatures. Now, an accumulator-heat exchanger installed in the newest models allows the heat pump to perform efficiently in temperatures as low as -30°C (-22°F).

Heat pumps cost around $2,000 to $3,000, but because they can take the place of air conditioners as well as heaters, they are a good investment.

The Thermostat

The control center for your heating system is the thermostat; for major energy savings, you should really have several thermostats throughout the house, either one in every room or one in every temperature zone of your house. This will allow you to keep some rooms, especially those you use only part of the day (or night), at lower temperatures, thereby saving you fuel and money.

In general, though you will save substantially—15½ to 20 percent—if you keep the thermostat at 18°C to 20°C (65°F to 68°F) during the day and 13°C to 16°C (55°F to 60°F) at night. Clock thermostats, which have timers, will automatically raise and lower the thermostat if you don't remember to do it yourself.

Once you've hit on your preferred temperature level, don't tamper with the thermostat. Constant adjustments make the heating system work too hard and waste fuel.

If you are accustomed to 24°C (75°F) most of the time and decide to turn your thermostat down (which you really should), do it bit by bit, not all at once. Going through the procedure slowly will give your body time to adjust to the lower temperatures and still feel comfortable. Turn it down a couple of degrees at a time, allowing two or three days between adjustments until you reach a good comfort level—which may be lower than you think! At this level you will probably need a sweater, and why not? This is actually healthier for you. Overheated, underhumidified spaces lower your resistance to infection. You may find that lower temperatures combined with adequate clothing and humidity make you less susceptible to colds.

Old Energy: Fire

Fireplaces, for all the notions of romance associated with their dancing flames and cozy glow, are the worst heat wasters of all heating systems. Fully 80 to 90 percent of the heat generated by a fire is lost up the chimney. Fire is sustained by the movement of air around the burning wood; the bigger the fire, the faster the air moves, more drafts are created, and the faster heat escapes.

Until Benjamin Franklin invented the Franklin stove, the fireplace was the only heating system known. In colonial times houses were built with massive central chimneys that serviced several fireplaces. Rooms were small and the size of each fireplace depended on its function in the room it occupied. The kitchen fireplace, which was also used for cooking, was the largest. Bedroom fireplaces were much smaller, because the size of a fireplace is dictated by the length of the chimney, and since most bedrooms were located on the upper floor of a house, the length of chimney was not adequate to support a large fire. But if the chimney were left exposed upstairs, some heat was radiated through the brick into those rooms to make them more comfortable—as long as the fire was going downstairs.

For most efficient fires, the opening of the fireplace should not be more than 12 times the opening of the flue or chimney.

If you are planning to build a new fireplace, consider its shape. A shallow fireplace with a curved back wall and splayed sides will direct heat out into the room. If you possibly can, place the fireplace on an inner wall where you can get maximum heat benefit from the warmup capacity of the chimney—just as in colonial days. If the fireplace is built into an outer wall, some of that stored-up "extra" heat will escape directly to the outside through the chimney's exposed three sides.

A couple of simple tricks will make your fireplace, old or new, work better. You can keep the damper partially closed or add a little chimney shelf at the bottom of the chimney, to reduce the flue opening, and thereby lessen the amount of heat escaping up the chimney.

Because the fire needs so much air to keep it going, you can pull in air from the outdoors, rather than indoors, by constructing a vent system. You'll have to make a hole in the floor or hearth in front of the fireplace, sized to hold a standard register cover. Under the hole, in the basement or crawl space, build a duct long enough to reach the outside wall. Tilt the duct slightly so that any moisture buildup will

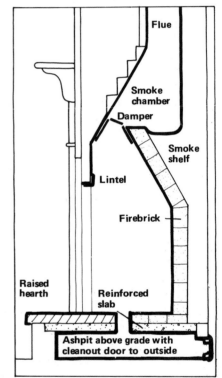

Sideview of a fireplace

Flue

Smoke chamber

Damper

Smoke shelf

Lintel

Firebrick

Raised hearth

Reinforced slab

Ashpit above grade with cleanout door to outside

Humidifiers

Winter air in the home is notoriously dry. Plants shrivel, doors shrink and your skin cracks as though the winter home is a desert. The dry air will steal moisture from everything—furniture, fabric and anything else, even your body. During summer, the moisture pulled from your body by evaporation helps to keep you cool; during the winter, you will feel distinctly chilly.

The most comfortable level of relative humidity (the amount of moisture in the air as compared with the maximum amount that the air could contain at the same temperature) during the cold months is about 35 percent. Most of our houses are not humidified to even half that level. Correct humidity will allow you to feel more comfortable with less heat. Most humidifiers come with

an automatic humidistat which constantly monitors the moisture content in the air. Humidifiers also come with a water-level gauge and an easy-to-fill reservoir; a well-insulated house will need about 3 gallons (11 liters) of extra moisture added to the air each day.

You can buy or rent humidifiers. Some are built for permanent installation and will add moisture to the whole house, while other models are portable and will handle one room at a time. Be sure to calculate the exact dimensions of the space you wish to humidify and select a machine of the right size.

drain to the outside. Make a hole in the outside wall to accommodate the duct pipe. On the outside a connecting elbow duct with a screen attached should point downward to keep insects, leaves and wind from entering your duct system.

Although andirons are very attractive, a hollow grate that circulates heated air and expels it into the room is a more practical solution to holding wood in place as well as generating warmth at the same time. This type of grate looks rather like a comb made up of pipes, curved to hug the wood and direct the air up and out. In combination with the floor vent, a hollow grate functions very efficiently to maximize fireplace heat output.

A heat exchanger, which consists of a vent and fan located in the chimney, will release some of the heat moving up the stack back into the living area. For obvious construction reasons, it is more difficult to install one of these in a brick chimney than in a prefabricated metal flue, but it is possible and worth the effort.

When the fire is dying down and you do not want to rekindle it, you can contain the heat—and even keep embers going until you want to relight the fire—either with a fireplace curtain or shield made of glass or asbestos, or with a snuff box that you can make from a noncombustible material.

When you leave the fire and it does go out, be sure to close the damper so that all the warm air will not go up the chimney.

Because the burning of wood leaves a gummy combustible residue called creosote in the chimney, it is essential to clean it out routinely.

Building a Masonry Fireplace

A masonry fireplace takes up a great deal of room. Because it is so massive, building one is an undertaking for new home construction,

---

No Heat—A Checklist and Some Remedies

1. Make sure that the switch that operates your furnace has not been accidentally turned off. All furnaces need electricity to operate.
2. Check the thermostat location. If for some reason the thermostat is being unnaturally heated (a strong ray of sunshine, an overheated lamp too close to the thermostat, a candle close to the thermostat) it may turn off the furnace automatically.
3. If the heat output is low but functioning:
   a. If you have a forced-air system, check to see that the fan belt is running, the dust filter is clean and the circulating fan is running. If the problem lies with the circulating fan, push the reset button; if there is no response, check the fuse or circuit breaker that controls the furnace.
   b. If you have a hot-water system, make sure the circulating pump is operating; push the reset button if need be. Check fuses or circuit breaker.
   c. If you have a steam heating system, the boiler may not have enough water in it. Let the boiler cool for about one hour, then fill it halfway.
4. When an oil furnace stops running entirely when the power is on and the fuse or circuit breaker is operating properly,

it may be out of fuel. Check the level of fuel with a dip stick, as gauges may be stuck or broken.
5. A gas furnace may stop running if the pilot light goes out. To relight the pilot, turn off both the main valve and the pilot-light gas valve. Wait a few minutes, turn the pilot-light valve on again and light it with a match. When the flame is burning smoothly, turn the main valve on again. All this should be done with the thermostat at its lowest setting.
6. Keep yourself warm with space heaters—portable electric heaters, camp stoves, catalytic heaters. Light a fire in the fireplace. If none of the above are available to you, gather in the warmest room of the house, one that may be heated with sunlight or that is on the southern side of the house, away from the wind. Dress warmly, eat warm foods and drink warm liquids. Gathering around the kitchen stove, especially if something good is cooking, will keep you warm. Exercise will also make you warmer—dance, run in place or jump rope.
7. If you live in an apartment and have no control over the heating system or the landlord, remember that most cities have emergency numbers to call if you are without heat. There are usually laws to determine how much heat your landlord should give you and at what times (see box on p. 35). Find out about these and your complaints may carry more weight.

8. If you are employing any sort of space heater, even a fire in a fireplace, make sure that there is adequate ventilation.
—E.L.

not for a renovation or remodeling project. It needs a solid concrete foundation at least 3 ft. (90 cm) deep and 6 in. (15 cm) wider on all sides than the actual fireplace to bear the weight of the fireplace and chimney. Multiple fireplaces using the same chimney require even greater bulk, and before you begin, check your local building codes to determine how deep and wide the foundation must be beyond the standard requirements.

Until this century, the hearth was the center of the home both figuratively and literally, and indeed, fireplaces are most efficient when placed in the center of the house. The chimney will give off radiant heat through the brick or stone and also offer the option of multiple fireplaces. Today, exterior installations are very popular, and such fireplaces make a good focal point in a living room or family room when no other fireplace is desired in the house.

Fireplaces should be of a specific size in relation to the room area they are to warm. About 5 sq. in. (32.5 sq. cm) of fireplace will warm each square foot (.84 sq. m) of floor area—a ratio of one to two. The standard fireplace opening is 3 ft. (90 cm) wide by 2 ft. (60 cm) high, but in general two-thirds to three-quarters the width. Most fireplaces and chimneys are constructed of brick or stone. They should be lined with special firebrick, and the flue with flue tile. Brick chimneys should measure at least 8 in. (20 cm) thick and stone at least 1 ft. (30 cm) thick; where a separate flue is installed, the brick can measure 4 in. (10 cm) thick around the flue.

Corrosion-resistant sheet metal, called flashing, must encircle the chimney at the roofline to prevent leakage, and a cap on top of the chimney will thwart rain, insects and rodents.

The first few fires you build in a new fireplace should be small ones, which will season the stones or bricks, and build up heat tolerance. If you begin with a big blaze, you may find cracks in the fireplace.

To build a fire in a masonry fireplace, select logs of well-sea-

---

Cleaning the Chimney

In order to prevent chimney fires caused by creosote buildup or soot accumulation, you must periodically clean out the chimney of a conventional fireplace and/or the flue of prefab, free-standing or wood-stove system. One simple way to slow the accumulation of creosote is to sprinkle salt over the ashes before each fire.

The cleanest way to clear out the soot is to climb up on the roof and work from the top down. First block all openings to the chimney—both the fireplace opening itself and any ducts for air intake or outtake or to a heat exchanger.

Fill a bag with rags and bricks to the dimensions of the chimney or flue and tie a rope or cord to the bag that is as long as the chimney is tall. Gently push the bag down into the chimney cavity and pull it up and down in a rhythmical manner until you feel you have dislodged all the creosote, which will fall down into the firebox. Then it is a simple process to collect the soot and throw it away.

If your flue is exposed, comes in sections and/or has elbows, you can disconnect the pieces and swab them out. Again, it is a messy process, so try to do the cleaning on a bed of newspapers or old dropcloths.

The Chimney Sweep

With a top hat and a bristle brush, the chimney sweep was a familiar figure on the rooftops of the Victorian age. He hasn't completely disappeared, and in fact is staging a comeback thanks to the high cost of fuel and a partial return to fireplace heat. The Chimney Sweeps Guild has 175 members, and some 850 others (including 7 women) have purchased special kits which contain everything right down to the top hat. If you are reluctant to tackle the soot, ash and creosote in your chimney, you can hire one of these professionals to do the job. They usually charge about $40 for the first chimney and $30 for each additional flue. To find the one nearest you, contact the Chimney Sweeps Guild,

---

soned, completely dry hardwoods which will burn more slowly with less danger of casting sparks. Oak, hickory and most of the fruitwoods are good. You'll need three or four logs, preferably split, measuring about 5 in. (12.5 cm) in diameter and between 16 in. and 22 in. (40 cm and 55 cm) long. If you use andirons, set them about a foot (30 cm) or so apart in the center of the hearth. A grate works more effectively than andirons to set up a circulating flow of air through the fireplace and back into the room.

Crumple up some old newspaper and lay them in the grate or between the andirons. Add kindling in a cross-hatch pattern and then place the logs on top in a pyramid, with the largest at the back. Roll up a single piece of newspaper and light it, holding it under the flue for a moment to set up a draft. Now light the crumpled newspaper under the kindling and step back and enjoy the results.

The best fires begin on a 1 in. (2.5 cm) thick bed of ashes. If you haven't built up any ashes yet, the fire may go out a couple of times before it takes. As the logs burn, turn them with tongs; after one of them is gone, replace it with a new log to maintain a steady, even heat from the fire. Contrary to what you may think, a really big fire is inefficient. It pulls too much air out of the room and loses heat up the chimney. Aim for a steady medium-sized flame and work at keeping it even. Pull up an old wing chair—designed to catch as much heat from the fire as possible—and pull out a book or let your thoughts drift as you contemplate the crackling fire.

Prefabricated Fireplaces and Woodburning Stoves

In the eighteenth century, Benjamin Franklin, recognizing the essential inefficiency of the conventional fireplace, invented a firebox made of cast iron which was placed inside the fireplace to make use of its chimney. This early "Franklin stove" burned better partially through wiser control of draft and partially by containing the fire behind doors.

Franklin stoves have their counterparts today in many different guises. Some versions are still made of cast iron, but most are constructed of easy-to-weld sheet steel. All have doors and all use a stove pipe to vent gases to the outside, either through the roof or through an outside wall.

One contemporary adaptation of the Franklin stove principle is the prefabricated fireplace. It is made of sheet steel and may be ordered ready-made or in kit form for do-it-yourself installation. Prefabs are boxlike, with pull-to-doors that contain the heat, metal flues to be crafted in section and cold-air intakes and hot-air ducts. Newest models have cold-air intakes that are damper-controlled ducts that reach to the outside, through the wall, to pull outside air in, rather than using the room air.

Like the Franklin stove, prefabs may stand free, or they may be installed through or against a wall and then veneered with brick, stone or any other noncombustible material. Because of the design of these fireplaces, a hearth is not necessary, although many people prefer to add one for looks.

Free-standing fireplaces, to be viewed "in the round," can be positioned anywhere in a room. They come in a great variety of sizes, colors and shapes. They may be spherical, conical, or boxy. They may be brightly colored or basic black. They may have pull-to doors and feature all kinds of exotic fireplace tools. There is one requirement, however, which they all have in common—they must be set on a noncombustible surface, such as brick, ceramic tile, quarry tile or pebbles in a tray. This hearth should extend about 3 ft. (90 cm) beyond the fireplace in every direction. The fireplace must also be positioned at least 30 in. (75 cm) from the wall, unless the wall is sheathed with a noncombustible material such as asbestos. Locating the fireplace too near a window may cause panes to crack because of the heat buildup.

Prefabricated fireplaces come with openings of three standard widths—28 in. (70 cm), 36 in. (90 cm) or 42 in. (1.05 m). Many models open to the front, but there are versions that open to the side for corner or island installation.

When you decide to put in a prefab, you can choose almost any location in your house, but you should try to avoid any interference at the roofline so that the flue can be installed straight up. If you want a certain location and find that you will run into joists, trusses or rafters, you can opt for an elbow installation or even a double-offset installation, which means that the flue will bend at certain points to get around any obstruction. With any elbow in the flue, some extra structural support will be needed. No flue can accommodate more than four elbows.

At the roofline, the flue termination should penetrate at least 2 ft. (60 cm) above the roof ridge, and should be higher than any nearby obstruction. The flue termination can be encased in a simulated brick chimney in a choice of colors and styles. This top housing can be fitted with a rain cap.

One version of the prefab is the circulator fireplace that uses intake and outlet ducts to divert warm air back into the room or into other rooms. Circulator fireplaces come with the standard openings, plus three additional sizes—32 in. (80 cm), 46 in. (1.15m) and 54 in. (1.35m). Adding fans to the ducts increases airflow. To augment the efficiency of the circulator, some models attach the intake ducts to the outdoors, making use of cold outside air rather than warm indoor air to create the fire draft. These cold air intakes may be located beneath the hearth, or in the sides of the firebox unit.

Gas-fired prefabs are constructed in much the same way as the wood-burning types. Gas is directed through a control valve, which may be located in the floor, side or back of the unit, connecting to the gas supply below the hearth. When you light the gas-fired prefab, an automatic control will open the damper and ignite the burner. Venting operates the same way as for wood-burning models. Safety shut-offs are provided, too. A removable hearth plate allows access to the gas supply, in case you wish to shut off the gas for any reason.

If you choose a gas-fired prefab, look for certification by the American Gas Association.

Electric prefabs are usually used much like space heaters, except that they are mounted on the wall. Electrics can only produce a firelike look. A heating unit is necessary to provide any warmth. This type of prefab is especially handy for mobile-home dwellers, for there is no extra equipment or hook-up involved.

Wood Stove. An alternative to fireplaces, whether traditional or contemporary, is the wood stove. In the nineteenth century, wood-burning stoves were the major source of heat used for cooking. These woodburning cookstoves are still available today, but more prevalent are the simple heating kind. They may be made of cast iron, but more often are made from steel, with cast-iron doors. When selecting a wood stove, it is wise to check the thickness of the steel and determine how it is welded. Thin steel will bow out when hot, and therefore the

The Wood Stove

A wood stove is a great thing. On a winter day, there is nothing like it. There is always a kettle boiling, bread rising, soup simmering, and boots drying around it. You can use the steady warmth to grow sprouts, make cheese and yogurt, dry herbs and fruits, or brew beer and wine. And when you come in from the chores, it's wonderful to stand by the stove and soak up the heat. We have other stoves in the house—principally, a large Ashley chunk stove, which runs twenty-four hours a day, September through May—but it's the kitchen stove that's the heart and soul of our existence, and we love it.

My first wood cook stove was a warped old Kitchen Queen, some seven years ago. When I think back on those days, and my ignorance, it's a wonder I didn't burn the house down. I did in fact burn several large batches of bread, cakes, sauces, stews, and several times, my hands. In time, though, I discovered the oven controls, the drafts, and the ashes in the interior; I learned which woods to burn, and when to cut them. The day came when we could bank the stove (which is very small, as stoves go, and not too tight on top) and find three inches of hot embers there in the morning.

These days, many people are moving out to old farmhouses and rural areas. Many call it a simpler way of living, and there's no question but that its joys are simpler. Long walks, and visits, now and then, replace movies and television. Music is mostly homemade; clothes are individual, so to speak, to the wearer. Seasons change, children grow, and the steady rhythm of chores fills days that were once tense and uncertain. We grow closer to the earth, and life seems good, most days.

However, though our lives are less complicated, our roles in this world are not, necessarily, easier. An outhouse for example, is a very straight-forward sort of thing. But it takes more than the turning of a handle to keep it clean and safe and useful. Kerosene lamps are cheap and lovely things, but they need to be cleaned and trimmed and filled. So with a wood stove; you need to know a little about it, to enjoy living with it. What little I have gleaned from trial and error, I gladly pass on to you, in the hope that your rice never burn, your bread never fall, and your house be safe and warm.

The Fire Box and the Grates

The first consideration, naturally, is the fire box. This is the compartment in which the fire is built. The fire box should be lined with either cast iron or stove brick, cut in special shapes to fit the fire box, so the heat doesn't spread out in all directions.

Stove designs vary. Some load from the side, some from the front, and in some the top of the fire box lifts up in one unit on a hinge.

Under the fire box is a grate for ashes to sift down through. In many stoves this grate can be turned slightly by a removable handle. Turned one way, the grate is suitable for wood; the other, more open, way is for coal. When such a grate is set for coal, it has wider spaces between the cross pieces, because coal tends to build up a very solid bed of clinkers and ash; you always want there to be some draft. The wood grate is tighter, to prevent the embers (which are necessary for a good fire) from continually

falling through. It is impossible to really bank a wood fire over a coal grate. Coal grates and coal/wood grates are more common in coal country. If your stove is in a Vermont farmhouse, it is probably a wood grate. Sometimes grates are built so that they can be turned slightly, in order to allow you to sift down ashes without getting your wrist black. If the handle is missing, or the grate doesn't turn, use a poker—or, if the stove is cold, a small stiff brush is handy.

The ashes underneath must be removed every few days, for which purpose it is handy to have a removable metal box under the fire box. Deposit these ashes outdoors, far from anything flammable. If your ashes are *all* from hard wood, you may use them to make soap. Ashes are also very useful in the garden mulch pile. If you have an outhouse, you can dump a scoop of ashes down the hole once a day to keep it from getting smelly, but set them aside first for twenty-four hours to make sure they contain no live embers.

On the side (and sometimes the front) of the stove, you will find various sliding or hinged doors, which are called drafts. When opened, they let in a draft of air, making the fire burn hotter. When closed, the fire will be "banked"—it will burn more slowly, and for a longer time.

The equipment for your stove should include a handle for lifting the top pieces out and moving them around. When you use it, hold the circle of cast iron at an angle, so it won't slip off and go clattering on the floor, your foot, or small creatures below.

The Oven

So much for the more obvious features. The next thing to consider is the oven. In order to look at the oven closely, change into your least perishable clothes, and arm yourself with a flashlight, a dust pan, brush, and wads of newspaper. Dismantle the top of the stove and take a look inside.

At the center back, where the stove is hooked up to the stove pipe, you will find a small sliding door. This has a control that enables the door to be opened and closed without opening the top of the stove. When the door is open, the oven is OFF. The heat and smoke from the fire will simply go through the open door and up the chimney. Any attempt to bake with the oven off will result in food baked on the top and one side, if at all.

Slide the door shut. The heat and smoke should go across the top of the stove, down the side, along the bottom and up the back—and out the chimney, through a passage up the back. In some models it goes down a divided side, around the bottom in a U-shaped passage, back up the side and out through a passage in the top. The general idea is the same: heat will surround the oven and bake your goodies from all sides. It will, of course, be hottest on the top, since the top is nearest the source of heat. For this reason I like to leave a light layer of ashes on top of the oven when I clean it.

Periodically, the inner workings of the stove become clogged with ashes, and you should clean them out. There is a simple tool designed for this that makes the job very easy; it consists of a length of stiff wire about two feet long with a small square iron attached to the end at a right angle, like a hoe.

With it, or some such device, you first sweep the ashes off the top of the oven into the fire box and shake them down into the ash box. Then, under the oven (either on the front or side of the stove) you will find a small removable panel that enables you to

Working Parts of the Wood Stove

Most wood stoves are constructed inside roughly like this:

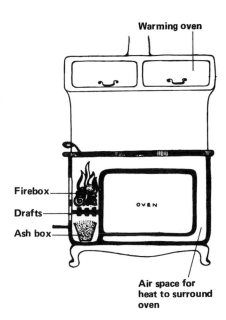

Warming oven

Firebox

Drafts

Ash box

OVEN

Air space for heat to surround oven

get at the ashes under the oven.

Scrape the ashes out carefully onto a newspaper, inspecting from all angles with a flashlight to make sure all the passages are clear. Replace the panel tightly.

You should not have to clean ashes out of the oven liner very often. In the winter, when I run the stove every day, I do it about once a month. In the summer, I do it maybe once. A lot depends on what you burn; the great stove-clogger isn't wood ashes at all but paper ashes. If you dispose of your paper garbage elsewhere your stove will stay clean a lot longer.

While you're at it, you might as well invest in a bottle or can of stove polish and blacken the cast-iron parts of the stove. Stove polishes are mostly combinations of grease and charcoal. You rub them in, wait ten minutes, and polish off the residue. Most of them work a lot better if you let them sink into the cold stove for a few hours before running it again. What brand you use will probably depend on what's available. In a pinch, you can just rub in a little vegetable or mineral oil. The important thing is to keep the pores of the cast-iron oily so that it doesn't soak up moisture and become rusty. I usually polish the stove once a week in winter, and once a month in the summer. Remember to oil or polish the stove thoroughly if you are going off on a long trip, since more moisture will be in the air while you're gone.

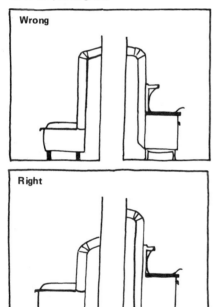

Wrong

Right

Stovepipes

The next thing to consider is the stovepipe. Most stoves are hooked up to the chimney with lengths of stovepipe. If you are putting in the connections yourself, keep the system as simple as possible. The more elbows and/or horizontal footage you have, the harder it will be for the smoke to be drawn up the chimney. Also, if you have more than one stove using the same chimney, they shouldn't be connected to the chimney at the same level.

A stovepipe must be fastened together with sheet-metal screws, so it won't fall apart. The vibration of wood stoves is slow but sure, and sooner or later the pipes will fall down. Stovepipes should also be securely cemented into the chimney with stove cement (a special concoction that comes

apart easily for stovepipe cleaning). The fine cracks around joints can also be sealed with this stuff to insure a good tight connection. If you have local natural clay, this will work just as well (in fact, better) on the connection to the chimney, but it won't work between the pipes. Properly fitting pipes will not need cement, anyway.

Periodically (once or twice a year) you ought to pull down the stovepipe and clean it out with a wire brush. Creosote will not collect in it as much as in stovepipes from heating stoves, but it has been known to accumulate and catch on fire in houses where much soft wood is used. More on creosote under "Chimneys," below.

Your stovepipe should have one or more handles on it. These are the "dampers" and they control the speed at which hot air is sucked up the chimney. Inside the pipe is a flat circle of light metal that can close off, open, or partly open the stovepipe.

When the damper is all the way open, the fire may burn too fast for your purposes. You will have to experiment with the drafts and dampers to find the combination that provide the right amounts of heat for your purposes.

If your stovepipe doesn't have any dampers, you certainly ought to get at least one, and install it high on the pipe. Better yet, get two, and set them about 18 in. (45 cm) apart. With one, you will be able to keep a fire going for hours, and since it keeps the heat from shooting up the chimney, the oven will stay hot for much longer. With two dampers, you will probably be able to bank a fire so tightly at night that there will be hot coals in the ashes in the morning to start a new fire with. In any case, the stove will still be warm. Dampers cost less than a dollar, and all you do to install them is drill or punch two small holes in the pipe.

Chimneys

If you are moving into an old house with a wood-burning stove, one of the first things you should do is inspect the chimney. Check it for leaks and loose bricks; a surprising number of old chimneys are in various stages of deterioration. A leaky chimney will draw air through the cracks instead of through the stove, making the stove burn poorly.

Occasionally, after long disuse, chimneys become obstructed by bird or rodent nests. Ashes collect in the bends and in the bottom. The real problem to watch out for, though, is creosote. Creosote is a sticky, black, tarlike substance that collects in chimneys and stovepipes when smoke from creosote-laden wood comes into contact with a cold chimney. It is highly flammable.

The wood with the most creosote in it is pine, but hemlock and other soft woods have plenty of it too, especially if it is "wet" or freshly cut from a living tree. If the chimney is clean to begin with, and you do not burn soft or wet wood, you should not have much problem. In the far north, though, where hard wood is sometimes scarce, people do burn it. They have various ways of cutting down on the danger. One is to have a very thick central chimney, rather than a thin one on the side of the house. A central chimney will stay warmer and hence less creosote will collect. And, of course, if the chimney is thick less heat will be conducted to the wood of the house itself in case a fire occurs in the chimney.

When you first move into a house, and once a year thereafter, you should clean the chimney. One way to do this is to get hold of a nice big bunch of chains, tie them up in

a burlap bag, and lower them down the chimney on a clear, still day when the stoves are out. Or you can trim all but the top branches of a soft wood sapling and use it like a giant bottle brush. To clean ashes out of the bottom there is usually a loose stone or brick at the base of the chimney.

This will generally do the job, but if you really have a big creosote problem, and a nice heavy central chimney, there is another way to deal with it. You can shove a batch of lighted newspaper in the chimney and burn out the creosote yourself. You should, however, choose a very still day for this project, a day with snow on the roof, and collect some friends to help you keep an eye on the fire. This is a very radical measure but infinitely preferable to having an unexpected chimney fire at four a.m. on a windy night.

The sound of a chimney fire is a hollow roaring, but sometimes you can't hear it, especially if it's windy outside. If the creosote is burning near the top, you might see flames shooting out. The one time we had a chimney fire, the only sign was a pervasive stench that came drifting down the stairs . . . which turned out to be from rodent nests (and worse) packed tightly against the cozy chimney, inside the walls. I shut all the dampers in all the stoves and it went out. There wasn't much creosote, I guess.

If you have a chimney fire you should really call the fire department, but not everyone has a phone or a nearby fire department. Keep a very close eye on the house, especially the roof. Don't forget, it will take some hours for the heat to work its way through the brick or stone. Keep water handy at all levels until you're sure the fire's spent itself and the chimney has heated up and cooled down.

Another problem people sometimes have with chimneys is that they are not tall enough. They should be higher than the peak of the roof, but sometimes even this is not enough; tall trees (that weren't there when the house was built) or even a strong wind will cause a "down draft" and smoke from the stove will billow out into the room, especially when you first try to light the fire. I have known people to deal with this by shoving a twist of lighted newspaper through the stove into the stovepipe in order to get the draft moving upward. I think this is a dangerous practice, though. If you have to go to such extreme measures, there is something amiss with the system. It may be only a matter of rearranging the stovepipes so there are less bends; but it may also be that your chimney needs to be lengthened. Unless you know a fair amount about masonry, better get somebody who knows how to lay bricks or stones to lengthen it. Nobody wants a loose brick sliding off the roof unexpectedly one day. In any case, don't add stovepipe to make it higher, as creosote will collect in it very quickly. And don't put a stovepipe cap on top of it. Caps slow down the draft and the longer the smoke lingers in there, the more creosote sticks to the inside of the chimney.

—Susan Restino

welding should be slightly convex, and free of pores.

These days, wood stoves are designed to be airtight, so that the wood will burn slowly and completely, leaving little ash. Such burning makes the stove a more efficient heat producer than the Franklin type, but it also creates increased creosote buildup in the flue, making routine cleaning a necessity to insure against a chimney fire. There are several varieties of airtights. Some draw air in over the wood, some under and some both over and under. Versions with two drafts, developed in Scandinavia, are the most efficient, as the wood burns from front to back, and remains even.

Boxy "circulators" depend on their thermal efficiency by means of a warm-air sandwich between the firebox and an outer metal jacket. As the air warms up in this space, it rises and moves out into the room.

One way to complement any wood stove's heating power is to add a heat pipe or duct in the chimney pipe near the ceiling or on an upper floor. That way, some of the escaping heat can re-enter the house. Exposed chimney pipe, too, will give off heat on its own, by radiation.

New Energy: Solar

The combination of escalating fossil fuel prices, dwindling fuel reserves and specific government endorsement for research and development has made the idea of using energy from the sun very attractive. It is, of course, far simpler to assimilate a solar system into the designs for a new house, but there are certain solar heating systems—especially one for heating water for domestic use—which can be "retrofitted" or adapted to an older home.

Solar energy is not entirely new. Individual applications for heating houses have been around for over a quarter of a century, but mass-production of solar components is a fairly recent phenomenon, and solar technology is currently developing rapidly. Even so, some of the most successful applications of solar power are simple, inventive and use little or no machinery. Purists prefer these systems, for they can be personalized in any number of ways and are wholly ecological, destroying or wasting nothing to produce the energy a house requires.

All are so-called passive systems. They make use of the bulk of the building mass, plus glass or plastic, and nature's own heat exchangers cum insulators: earth, air and water. Passive systems most often operate on a "greenhouse" principle, making use of great expanse of insulated glass on the south-facing wall of the house, usually combined with inner masonry masses or walls painted black and a "wall" of air between the two. The air sandwiched between the glass and masonry is heated by the sun; the heated air rises and passes into and through ducts. Then it is distributed simultaneously into the house and to a storage area or bin that is usually filled with rocks. The rocks absorb and retain heat, and when the outside air is cooler, especially on rainy or cloudy days and at night, heat is released into the house via the ducts. The ratio of rays to heat in the passive solar system is as follows: 1 sq. meter (1.2 sq. yd.) of insulated glass draws in enough heat to maintain 10 cu. meters (13. 5 cu. yd.) of interior air at a temperature that is comfortable. The passive system should provide at least 35 percent—and up, depending on climate—of your heating needs. For materials, installation, and the requisite storage bin, you would pay around $1,500 to be properly outfitted.

There isn't space to.discuss fully all the components of a solar heating system, but we can mention some specifics. The classic air wall was developed by Felix Trombe, a French architect. Called the Trombe solarwall, the system is simple in concept. Heat is built up in the air layer between a wall of insulating glass and a wall of masonry painted black. The masonry wall can be permanent, but this blocks out light and landscape views. Instead, movable panels of a lightweight insulating material painted black on the outer side and reflective on the inner side allow flexibility when light and landscape are desired. At night, the panels are pulled in to contain the heat within the house.

There are many designs and variations of insulators. With glass and little or no masonry backup, the house itself becomes a green-house—without humidity. Heat bakes through the glass during the sunny hours, and must be contained on cloudy days or at night by one or another means of insulation. Of these, Beadwall is unique. It was designed by Steve Baer of New Mexico for use with passive solar heat-ing systems and consists of a glass sandwich. At night, polystyrene beads, pumped in between the two panels of glass, hold in the heat collected during the daylight hours. In the morning, the beads are sucked back into their storage chamber.

Another Baer invention is the Skylid, which acts like a louver in the roof, opening and shutting to let in or trap solar heat. Skylids function without a power source other than the rising and falling hot air.

As an alternative to air and the air wall of whatever design, water can be the heat-inductive medium. For his own house, Steve Baer created a honeycomb grid which holds a series of 55-gallon (200 liter) drum barrels, filled with water and painted black on their sun-facing ends. The water heats up during the day; at night, wall-sized panels on the outside of the house lift up over the wall to trap the heat. These panels are contructed of insulation foam sandwiched between aluminum sheets. During the day the aluminum panels lie flat on the ground next to the house, serving as a highly reflective medium, beaming intense sunlight against the barrels.

A rooftop water system was designed by Harold Hay of California. It consists of large plastic bags of water resting on a flat metal roof. To absorb or store heat, insulating panels expose or cover the bags, called solar ponds.

Baer's and Hay's simple systems work as well as they do not only because of the creativity of their inventors, but also because of the climate zones in which they are used. New Mexico and California are generally sunny and dry, conditions most conducive to solar power.

Most temperate zones need some storage system other than the actual collectors. Pebbles between 2 in. and 5 in. (5 cm and 12.5 cm) in diameter are the best medium for storing hot air. Water, one gallon (3.7 liters) in ratio to 1 sq. ft. (930 sq. cm) of heat space, is also an effective conserving medium for heat.

Solar Heating

If you are interested in solar energy systems for heating and cooling, the place to contact is The National Solar Heating and Cooling Information Center (NSHCIC), a private research facility working for HUD and DOE. Someone there can answer most questions about solar energy, from where to go to find it in your area to the federal and local programs that are willing to provide loans or other assistance for solar installations. The address is:

National Solar Heating and Cooling Information Center,
P.O. Box 1607
Rockville, MD 20850.
They also have a toll-free telephone number: (800) 523-2929

(in Pennsylvania: 800-462-4983). The NSHCIC also publishes a rundown of state legislation concerning solar energy systems. This comprehensive list is expanded and brought up to date regularly.

—A.R.

Active solar heating system.

This is much more complicated than any of the passive systems we've been talking about because every functional part is mechanized. Such a system also costs anywhere from $4,000 to $10,000, depending on climate.

The most important component is a collector. Usually installed on the roof and angled toward the sun, the collector acts like a heat-absorbing sandwich. The outer layer is composed of double-glazed insulating glass and the inner layer is metal, often copper or aluminum and usually painted black for greater heat absorbtion. The metal panel is backed with an insulating material so that the absorbed heat will not escape to the surface of the roof. If the active system is of the hot-air type, the heated air passes from the rooftop collector plate to the distribution system and then into the storage bin that contains rocks. If the system uses hot water, tubes that run through the collector plate will then move heated water into pipes and into a water-filled storage tank. Water will absorb more heat by volume than air.

As active solar heating systems become increasingly sophisticated in their design and efficiency, it appears that the liquid-type collector is the more attractive choice for home installation. With the major federal research and development grants going to large manufacturers and corporations, liquid-type collectors have been developed to the point where their manufacturers can actually include warranties and guarantees of their maintenance and servicing. Before you decide on a collector system for your own house, be sure to check the performance test data and the rate of efficiency of the collectors in your climate zone and choose the most efficient model.

Are there advantages of liquid versus air collectors? Liquid types are susceptible to corrosion while air types are not. However, many manufacturers are improving the durability of liquid collectors by designing them with a corrosion-resistant metal, such as copper. Liquids must be mixed with an antifreeze solution so that they will function in cold climates and so that the collectors will not be damaged by freezing.

Still at the experimental stage are concentrator collectors or solar discs which concentrate the diffused rays of the sun into higher intensities thus producing much more energy than is now possible with current solar collectors. Solar collectors will one day be used to cool the house as well as heat it.

Whichever collector type you choose for either an active or passive solar heating system, it should be able to sustain a temperature of at least 60°C (100°F) over the temperature of the outside air. This temperature differential will insure an indoor temperature range of 30°C to 50°C (85°F to 120°F), about the same as that produced by a heat pump. This temperature will feel colder than the heat produced by a conventional furnace by about 10°C (20°F), but it is a comfortable temperature once you get used to it.

If you decide to install an active collector, be sure to check your local building codes and insurance policy. There may be restrictions on what you can construct and you also want to be sure you are adequately protected by your insurance.

Because solar energy shows the promise of low cost for a high return, and is an inexhaustible source of heat, there are tax incentives —up to $2,000 as a write-off if you go solar from the start. Also, more and more solar manufacturers will be able to work in guarantees and maintenance to their advantage as well as yours, making an investigation and possible investment into the new frontier of solar energy well worthwhile.

Solar furnaces

Many older houses are designed and sited in a manner that is not conducive to installing rooftop collectors. An alternate solar source is the solar furnace. It stands separately in a sunny spot near the house, and direct heat collected from the sun is introduced into the house via a simple duct set-up.

There are two types of solar furnaces currently on the market. One is a small A-frame structure, the other a basic lean-to. Both are faced with collectors that must be oriented towards the sun. Inside, rock beds soak in and store the heat, which is then drawn into the house through the ducts by convection and with blowers. A cover, cloaked on its inner surface with a mirror, tilts out on sunny days to refract the sun's rays with greater intensity into the furnace. When the sun isn't out or at night, the cover fits over the collector face to retain the accumulated heat.

Prices for solar furnaces run between $3,000 and $4,500, installed.

Creature Comforts

Curling up with a favorite book, sipping hot cocoa in front of a blazing fire, or baking cookies and cakes in a steamy kitchen—we all have special ways to warm up and feel cozy during a long cold winter. Some methods are based purely on psychology; and, indeed, the illusion of warmth does promote the actual feeling of warmth. The colors we wear, the furnishings with which we surround ourselves, and the friends we share our time with warm our spirits and can help to warm our bodies. Feeling content and relaxed helps to keep us warm. One of the bits of advice frequently given if one finds oneself stranded in winter is to avoid succumbing to fright and loneliness; if instead one can spend the time appreciating the beauty of a woodland or a snow-capped mountain, the frigid temperatures might be easier to bear.

That, of course, is an extreme case. But for most of us anticipating another gelid winter, some warmth might be derived from the relaxed sense of knowing we are prepared for what is to come. In the pages that follow some of the practical aspects of preparing for personal comfort will be explored. What clothing to buy, and how to wear it is of major significance. There are ways you can decorate your rooms to supplement the insulation and heating your home already has. Knowing how to keep yourself toasty during a winter night, even if your thermostat is turned way down, can do a lot to increase your comfort. And among all these practical ideas and solutions, you will find some "illusions" that you may wish to try to warm the spirit as well as the body.

What Makes Us Feel Cold

Unlike our nearest relatives in the animal kingdom, we humans do not come equipped with a furry outer covering to keep us warm. We must, nonetheless, maintain a body temperature of approximately 37° C (98.6° F). This average temperature represents the balance between the heat the body produces and the heat it loses.

Your body heats itself by metabolizing the food you eat. The heat is generated in the torso area and is carried to your extremities through the bloodstream. It's an efficient system until external temperature begins to drop. When that happens the blood supply near your skin surface is cooled by the air around you at a rate greater than normal. This upsets your body's heat-production/loss balance, and, as a result, you feel cold. Your body has a sort of thermostat, the part of the brain called the hypothalamus, which reacts to the cooled blood passing through it and signals your body to reduce heat loss and increase heat output simultaneously, so vital functions will not be impaired. Because the vital organs are located in the torso and head, these areas get priority; warming blood is concentrated there. Hands and feet are thus the earliest victims of cold. The blood vessels to the extremities are constricted to reduce radiation. Blood flows through the constricted vessels at a slower rate, giving off more of its oxygen than usual, and a bluish tint may appear on the surface of your skin. (This happens to all of us to a degree, but smokers, drinkers, and the elderly will find that their extremities chill even faster than those whose circulatory systems are in top form. It is worth paying attention to this if your hands and feet often feel colder than the rest of your body. Feeling cold is *being* cold; sufficient supplies of warming blood are not getting to your extremities, and you must compensate for this by clothing yourself against the cold.)

At the same time, your arm and leg muscles tense up and shiver, increasing the pumping action of your heart, which in turn increases the heat output of your body. "Goose pimples" also result from the tightening of the muscles; but this mechanism, which served to fluff out our hairy ancestors' natural fur coats and thereby to increase the insulating qualities of same, does little to help the relatively smooth-skinned human.

But if evolution has deprived us of built-in insulation, our intelligence and our extraordinary ability to adapt has permitted us to develop external means for retaining the heat our bodies produce. Beginning perhaps with a pelt slung around the shoulders, *Homo sapiens* learned to use nature's resources to create adequate covering. As early as 5000 B.C., cotton was being cultivated and used for clothing in Mexico, and wool garments were being worn in Babylon (literally the "land of wool") in 4000 B.C. Today we have an overwhelming choice of fibers and fabrics, some better suited than others to retaining body warmth. How warm a particular garment will be depends on the fibers used, the construction of the fabric, and the

---

Hypothermia

What It Is
The lowering of body temperature below 37° C (98.6° F)

Where It Strikes
The whole body

When It Strikes
Whenever the body loses its interior heat to the environment. Hypothermia can be caused by prolonged exposure to very low temperatures in the winter, by rapid chilling during a summer rainfall in the mountains, or by falling into cold water when boating in any season.

Symptoms
Skin grows pale.
Uncontrollable shivering and stiffness; body feels painfully cold and movements require an effort of will.
Dizziness and drowsiness occur.
Sight becomes dim.
Victim feels disoriented.
In time, shivering stops and movements become impossible.
General numbness sets in.
Lightheadedness or nausea may occur.
Real danger point is when body feels warm, and victim is overcome by drowsiness.

What Is Actually Happening
Critical period is when body temperature is between 17°- 33° C (82°- 91° F). Heart may fibrillate in an attempt to pump more blood; heart failure is common result.
If victim survives the critical period, heart rate and respiration rate may be severely curtailed. Eventually, the tissue and cells are destroyed as the body freezes.

Treatment
1. Initiate artificial respiration or cardio-pulmonary resuscitation if victim is unconscious.
2. Bring victim to a warm environment, or, if necessary build a temporary shelter and a fire.
3. Warm the victim as rapidly as possible, either by covering him with blankets or placing him in a warm bath.
4. Remove any clothing that may be constricting the flow of blood.
5. When consciousness returns, treat as for shock.
6. Give victim warm liquids.
7. Make sure that the victim is totally dry if he had been placed in a tub; keep him in a warm bed, covered with blankets.

Prevention
Make sure you are in good physical condition if you plan to spend long periods outside, even if you're just shoveling snow.
Eat properly, and take extra quick-energy food along for long outings.
Dress properly.
Build a shelter from wind and weather if you are stuck outside.
Avoid panicking if you are lost.

finishes applied to it. And it can all become impossibly confusing. Here are a few basic principles to make it easier to answer the all-important question: "But will I be warm enough when it's really cold out?"

Dressing to Keep Warm

The key to keeping warm is insulation. Unless you plan to wear an electric blanket with the world's longest extension cord, the warmth you feel from your clothes will be your own body's warmth; the service performed by your clothes is the retention of that warmth.

Still air provides the best insulation; if a fabric allows air to move through it, warm air from your body will flow toward the cooler air around it. The insulator is most effective when it traps air in tiny separate compartments. The loft of a fabric—its ability to spring back into shape after it has been compressed—is one good way of evaluating insulation potential, since resilience is a sign of capacity to entrap air.

But how well your garment traps air is not the only criterion, just as air temperature is not the sole determinant of how cold it feels. If the air outdoors is moving—mild breeze or savage gale—the flow of warm air away from your body will be greater than if the air were perfectly still. How much greater depends on how fast the wind is blowing. That's what the wind-chill factor is about. How well the fabric resists the wind is another important determinant of how warm the garment will be.

Aside from cold and wind, the winter dresser must be concerned with moisture. As we know, air is a poor conductor of heat; it so happens that water conducts heat very well. This means that in cold weather, moisture will conduct your body heat *away* from you and toward the cooler air around you. The direction of movement is always from the warmer to the cooler, because nature loves equalizing things. Whatever the physics involved, the net result is a cold, clammy feeling that is very unpleasant indeed.

Moisture produced by the body is as much of a problem as moisture in the atmosphere. Regardless of how cold it is, if you dress heavily and engage in even moderate activity, you will begin to sweat. If that moisture cannot escape from your body and evaporate through your clothes, you will feel cold. And, of course, if your clothes are soaked by rain or snow, the same problem exists. Furthermore, the

---

The Wind-Chill Factor

If you listen to the weather reports, you've heard the expression **wind-chill factor.** Perhaps you have ignored it and focused your attention and outdoor plans on the temperature alone. If you have, you are making a big mistake. In fact, it is the combination of air temperature and wind speed that determines how cold it is outside—and it's not just how cold it **feels,** but how your body can be affected physiologically. That is to say, it is entirely possible to suffer from exposure when the air temperature is not extraordinarily low if there is a stiff wind blowing. For example, a 30-mph wind combined with an air temperature of -7° C (20° F) will for all intents and purposes make it feel like -28° C (-18° F). And under those conditions, frostbite and hypothermia are frighteningly possible.

Wind Speed in MPH				Temperature Centigrade/Fahrenheit
0	10	20	30	
4°C/40°F	-2°/28°F	-8°C/18°F	-10°C/13°F	
-1°/30°	-9°/16°	-15°/4°	-19°/-2°	
-7°/20°	-15°/4°	-23°/-10°	-28°/-18°	
-12°/10°	-23°/-9°	-32°/-25°	-36°/-33°	
-18°/0°	-29°/-21°	-39°/39°	-44°/-48°	
-23°/-10°	-36°/-33°	-48°/-53°	-53°/-63°	
-29°/-20°	-43°/-46°	-55°/-67°	-62°/-79°	
-34°/-30°	-50°/-58°	-63°/-82°	-70°/-94°	
-40°/-40°	-57°/-70°	-71°/-96°	-78°/-109°	

insulating ability of any fabric is decreased when it is water-laden, since the spaces where air might be trapped are filled instead with water, which, in turn, is busily conducting heat away from your body. Another danger of wet clothes is that the body may work too hard to increase its output of heat simply to dry the clothing. Moisture produced by the body must somehow be drawn off; moisture from an external source must be eliminated.

The ideal winter fabric, then, would have to be a superior insulator, with lots of space between fibers to trap still air; but wind resistant, its fibers tightly woven to keep air from passing through; impermeable to water, with few, if any, pores to permit moisture to enter from outside; but breatheable, so dampness can evaporate from the body. Obviously, such a fabric exists neither in nature nor in the laboratory. Trade-offs have to be made, and choices must be determined by the sort of weather you plan to encounter. Some fabrics come closer than others to being all things for all conditions. Let's look at what's available.

What Fabrics Keep Us Warm

The Naturals
Wool emerges as the most versatile winter fabric. Wool fibers have a natural curl or crimp, which makes each fiber stand away from the others, thereby adding to the bulkiness or loft of a wool fabric. The insulating qualities of woolen fabric can be further augmented by napping, a process that raises short fibers to the surface by rubbing the cloth with wire brushes. (Napping is part of the manufacturing process; it is not something you do at home.) The napped fibers of various lengths create more air-retention pockets.

Woolen yarns can be woven tightly, which makes for greater wind resistance. In addition, the cloth can be finished by fulling, a process that makes the fabric more compact and thicker by submerging it in warm water or weak acid to shrink it 10-25 percent. Melton cloth, for example, is heavily fulled, making it an effective windbreaker. Military uniforms (including the navy pea jacket), coats, and snow apparel are frequently made of Melton cloth.

Wool is naturally water repellent and will stand up quite well to moderate rainfalls. It can be chemically treated to further increase this quality. What makes wool uniquely valuable as a winter fabric is

---

Waterproof/Water Repellent

Waterproof means absolute impermeability to water. Rubber, neoprene- or urethane-coated nylon or other woven fabrics, and vinyl plastic, for example, provide total protection against moisture from the outside. There are no pores in the material to let water in; similarly, moisture produced by the body cannot escape from the inside of a waterproof garment.

If you will be exposed to a lot of water for a relatively long period to time, you will probably want waterproof dress, but beware of the clamminess that can develop if your body moisture is trapped inside. Buy garments with ventilation holes and take care not to tuck waterproof pants into rubber boots, or waterproof sleeves into waterproof gloves.

Water repellent means a fabric will shed moderate amounts of water, will not absorb moisture to any significant degree, but will allow air to pass through it. Most fabrics can be chemically treated for water repellence, or they may be naturally water repellent, as is unwashed wool, which has natural oils that shed water.

Despite what some people believe, untreated nylon is not at all water repellent. It is hydrophobic, which means it will not absorb water into its fibers, but it doesn't shed water either. In fact, water will run right through untreated nylon, which makes it quick drying (an ideal fabric for bathing suits), but not at all useful when it is wet out.

Wool is the winter dresser's best friend, but it comes in many guises. Weight, weave, and finish vary and can make a difference in the qualities of insulation, wind resistance, and breatheability. Here's a bit of help for the wool-gathering shopper.

General Terms

Wool: refers to fleece wool used for the first time in the manufacture of a wool product; *new* or *virgin* wool.

Reprocessed wool: includes scraps and clips of woven and felted fabrics that are shredded to a fibrous state and then re-woven; the scraps are usually cuttings from factory workrooms, never from wool previously worn or used by the consumer.

Reused wool: old wool from wool products that have been used by the consumer; it is cleaned, returned to a fibrous state, and then blended into yarn for fabrics; also called *shoddy.*

Construction Factors

Woolen yarn: wool that has been carded to produce a loosely twisted yarn; it has short fibers, randomly arranged, that when spun will yield a soft, fuzzy surface. That quality will be retained in the weaving or knitting process.

Worsted yarn: carded wool that is further processed by combing to eliminate short fibers and to bring the remaining fibers into alignment. Worsted fabrics are more tightly woven and are smoother than woolen fabrics because of this difference in the yarns from which they are made.

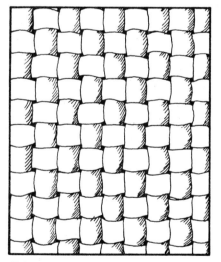

Plain weave: the most basic form of weaving, it is the simple over/under interlocking of horizontal (weft or woof) threads and vertical (warp) threads; the interlock is at a 90-degree angle.

Twill weave: The warp and weft threads interlock at a 45-degree angle in this weave, and the resulting fabric is characterized by diagonal lines in the weave. Twill can be made thicker and denser than plain woven fabric because the warp threads can be set closer together for twill threading.

Felting: A fabric-making process that does not involve weaving; wool fibers are treated with heat, moisture, and pressure so that they mat together permanently to form a fabric.

Knitting: Another alternative to weaving in which loops of wool yarn are interlocked together to form a fabric.

Some Popular Wool Fabrics

This is a partial list of some of the more common names under which wool fabric is available, particularly those fabrics best suited for cold weather wear.

Challis: a lightweight, plain weave woolen, primarily used for children's and women's indoor clothing.

Chinchilla: a heavy twill-weave napped fabric that resembles chinchilla fur, used for coats and jackets.

Felt: a nonwoven fabric, used in coats, hats, boot liners.

Flannel: a catch-all term for various weights of woven, napped fabrics, made with either woolen or worsted yarns, and used for shirts, coats, suits, pajamas.

Gabardine: a twill-weave fabric known for its durability; the surface has fine diagonal lines. Frequently used for coats, suits, and uniforms.

Homespun: made of coarse yarn, rather loosely woven in plain weave. Used for coats, jackets and suits.

Jersey: a knitted fabric used for dresses, blouses, and suits.

Mackinaw cloth: an extra-heavy cloth, napped on both sides, used for jackets.

Melton: a heavy, fulled twill-weave cloth with short, clipped nap, used for uniforms, snow wear, coats, and jackets.

Serge: a twill-weave cloth, known for its durability; used for suits.

Tweed: a rough-surfaced fabric, usually a twill weave, light in weight but durable. Used for suits, coats, jackets, and hats.

that whereas it repels water from an outside source, it can absorb as much as 30 percent of its own weight in perspiration. This is because the fibers can absorb water vapor and then slowly release it to the atmosphere. As an added benefit, as moisture combines with wool proteins, a small amount of heat is generated.

Although cotton does not have the versatility of wool, its basic properties make it a welcome component of most winter wardrobes, and cotton fabrics can be constructed and finished in numerous ways. The best thing cotton has going for it as a winter fabric is that it is more comfortable than wool for such items as underclothing. Cotton is highly absorbent and will wick away moisture from the body. Although it feels damp sooner than wool does, it will also dry out more quickly. And like wool, cotton can be treated with water repellents so that it can be worn comfortably during heavy rain and snow.

Cotton can be finished in various ways to improve its insulating properties. A nap can be raised on it, as is the case with flannel and chamois cloth. And it is frequently used for pile fabrics such as velveteen, corduroy, and terrycloth. Pile fabrics are smooth on the inside and have short raised fibers on the outside. They feel soft and downy, and the extra fibers provide pockets to trap air. In corduroy and velveteen, extra weft accounts for the pile; in terrycloth, extra warp threads are used. Cotton can also, of course, be knitted. The loops and loosely twisted yarn characteristic of knit fabrics absorb more moisture, provide more air traps, and are more breatheable than comparable weights of woven fabric. These properties, and the fact that knits do not tend to cling when damp, as woven fabrics sometimes do, has been used to advantage in cotton thermal-knit underwear and socks. Knitted piles, such as velour and the plush cloth used for sweatshirts, are two other useful winter cottons.

Of the natural fibers, then, cotton and wool are probably your best bets; they are usually reasonably priced for the comfort and warmth they provide. Linen, a fine fabric for warm weather, draws heat away from the body; nor is it appropriate for undergarments, since it has a tendency to become clammy when wet. Silk, customarily an expensive fabric, offers some benefits to the warmth-seeker. It has the ability to absorb a great deal of moisture without feeling damp. This is especially useful where perspiration is a consideration, as in socks and underwear. Like a cotton sock, but more luxurious-feeling, a silk lining sock will wick away moisture from the foot to be absorbed by an outer sock.

Two other naturals deserve mention in any discussion of winter outerwear: leather and fur. Leather coats, jackets, and pants do have a place in a winter wardrobe. Although leather is not as good an insulator as textiles or fur—it simply hasn't got the loft needed—it does provide excellent wind protection. Leather is generally very durable and can be processed to be water- and stain-resistant. Even suedes can be treated with a protective coating. But naked leather will not withstand a drenching without getting stiff, so be sure the garment you buy has been properly treated. Although leather boots and shoes can be treated at home with a silicone coating for water repellence, home treatment is not recommended on other garments.

Fur, of course, was the original winter covering. Today it is a luxury item and strict prohibitions are placed on the use of pelts from endangered species. Politics and economics aside, fur is an excellent insulator worn on either the inside or outside of a garment. The colder the climate in which the animal has been raised, the warmer the fur. And a long-haired fur is usually warmer than a short-haired one.

The Synthetics

Try as they might, textile scientists have yet to come up with snythetics that equal wool as a winter fabric. But because so much of what is available to the consumer is either synthetic or natural fiber blended with synthetics, we should note some of their virtues and disadvantages. Blending synthetic and natural fibers will reduce the cost of an article of clothing made from the fabric, will often increase the durability of the item, and will make for easier care. On the other hand, wool mixed with synthetics will not be as warm as pure wool, and cottons blended with synthetics will be less absorbent than cotton alone. Although today's man-made fabrics are not as clammy to the touch as earlier attempts were, they simply do not breathe as well as cotton, wool, or silk. Acrylic (Orlon, Acrilan), the closest thing to wool, does provide high bulk and light weight; but because it lacks wool's natural crimp, it cannot achieve equal loft, and the fibers are more likely to flatten together and pill with wear, thus lessening its insulating properties.

The two most practical uses for synthetic fibers in a winter wardrobe are as lining and as the outer shell of outerwear. The slickness of a nylon taffeta lining makes it easy to slip on a jacket or parka over a bulky, roughknit wool sweater. Polyester fabrics with nap or pile are often used as linings in winter coats. They add an extra layer of insulation at a low cost. The fake fur look, inside or out, is achieved with polyester pile.

Nylon can be much more tightly woven than wool or cotton, making it an excellent windbreaker. It is far lighter than cotton, which makes it useful for sporty outerwear. Nylon can be made completely waterproof if coated with neoprene or urethane. Polyester fibers (Dacron, Fortrel, Kodel) are most often found blended with natural fibers in winter clothing to lower cost and increase durability and wind and water resistance. These blends are heavier than nylon, but they offer the breatheability of cotton as compensation.

Frostbite

What it is
Ice crystals forming within the skin cells.

When it strikes
When one has been outside in cold weather for a long time with inadequate clothing.
When cold temperatures are accompanied by strong winds.
When hands and feet are damp in very cold temperatures.
When clothing, especially footwear, is too tight, interfering with circulation.

Where it strikes
Fingers, toes, ears, nose, heels.

Symptoms
Skin may be slightly flushed at first, then turn white or grayish yellow.
Although there may be some pain initially, frostbite area will eventually feel numb and cold.

Treatment
1. Loosen tight clothing near affected area.
2. Cover affected area with clothes, blankets, scarves. If hands are frostbitten, place them under victim's armpits or between thighs.
3. Bring victim to a warm area as soon as possible.
4. Give warm liquids—*not* alcohol under any circumstances. Coffee is particularly recommended since it dilates blood vessels, thus aiding the flow of blood to the affected area.
5. Warm the frostbitten area rapidly with warm water (39°- 40° C, 103°- 105° F). Circulating or running water will be more effective than still water, and if you are using a makeshift warming vessel, such as a metal container, make sure the affected area does not touch the container.
6. When the skin begins to appear flushed, stop warming and begin to gently exercise the affected part. Some pain will probably be felt at this point; it is a good sign.
7. Dry area thoroughly and carefully and apply bandages. If hands and feet have been affected, place sterile gauze between fingers and toes to ensure that they do not rub against each other. *Do not* rub frostbitten tissue with ice or snow, nor rub or massage it with the hands or anything else. Frostbitten skin is very vulnerable to permanent damage from abrasion.

Prevention
Protect fingers, feet, ears, and nose with adequate clothing. Make sure that boot lacings are not too tight; loosen several times a day, especially in extreme cold.
Keep dry.
Exercise hands, arms, legs, and feet to keep circulation going.
Keep warm by a fire.
Do not smoke; this reduces circulation by constricting blood vessels.
Alcohol has the same effect.
If frostbite danger is significant, use frostbite prevention cream, which coats exposed areas of skin and retards heat loss.
Keep a close watch on your companions—victims of frostbite often do not realize that they are affected, but others can readily spot an overly flushed or exceedingly pale spot on the skin.

Keeping Warm by Layering

The insulating value of any fabric can be enhanced if it is used as a layer covering or sandwiched between other fabrics. It's the trapped-air idea again: each layer forms still air pockets with the layers above and below it. In this way, several layers of medium-weight, loosely fitting clothing will provide more warmth than a single heavy-weight garment. The principle has been long known to hikers, campers, and mountaineers; as the "layered look," it has hit the fashion scene in a big way recently. So, these days it is possible to dress for chic and comfort at the same time.

Layering makes it easy to adjust your attire to changes in temperature; if activity raises your body temperature, you can remove a layer. And if you have lowered the thermostat in your house, wearing a sweater is a simple, inexpensive way to compensate for a chilly indoor atmosphere.

But what is the best way to dress in layers? Let's start at the bottom.

Underwear

Perhaps the most popular form of underwear for cold weather is the thermal-knit type. Available as all-in-ones or two-piece sets, thermal underwear is usually made of cotton or cotton and synthetic blend with a waffle or honeycomb textured knit, which increases the fabric's ability to trap warm air. There are basically three types of knit construction used in winter underwear: circular, rib, and Raschel. Circular and rib knits tend to be the warmest; Raschel knit is somewhat stronger and will, therefore, stand up better to heavy wear.

Woolen long johns did keep our grandparents warm, but there is no denying they were itchy. Most people nowadays prefer cotton next to their skin; it does not itch and is good at absorbing moisture, which means you won't feel clammy even if you sweat a lot. (Incidentally, cotton tends to get more absorbent each time it is laundered.) But because wool provides the best insulation and greatest warmth, the itch problem had to be overcome. Enter double-layer bonded underwear. The layer next to the skin is usually 100 percent cotton, for comfort; the outer layer is a blend of wool, cotton, and nylon, for warmth, stretch, and durability; and in between is a welcome air layer, for insulation. Double-layer underwear can be found in either one-piece or two-piece sets and in a range of patterns and colors. Purists

---

Snow Blindness

What It Is
Not blindness at all, but a sunburn afflicting the tissues of the cornea of the eye.

When It Strikes
When the eyes have been overexposed to the bright glare of sun reflected off snow. Sunny days are most dangerous, but corneal burns can also be suffered when it is cloudy or foggy.
Symptoms may not appear until several hours after exposure.

Symptoms
Redness, burning, watering of the eyes.
A grainy or sandy feeling in the eyes.
Appearance of a halo around lights.
Headache.
Somewhat blurry vision.

Treatment
Protect eyes from further irritation. The most effective way to do that is to keep them closed so that the lids will not pass over the irritated cornea each time you blink.
Take aspirin to relieve eye and headache pain.
If burn is severe and pain persists, see a doctor, who will probably prescribe antibiotic ointment to guard against infection and will temporarily patch one or both eyes to reduce lid abrasion. It should be noted that like any sunburn, a corneal burn can be more or less serious. The severity of the burn will determine how it is treated.

Prevention
Take precautions when you are outdoors for any length of time in snow-covered landscapes. Wear a well-fitting pair of sunglasses or ski goggles. Wraparound sunglasses will protect your eyes peripherally as well as frontally.
If no eye covering is available, darken the area around your eyes with mud, charcoal, or soot to reduce the reflective glare off your own skin.
Make emergency sunshields from a strip of bark or cardboard. The strip should be wide enough to cover your eyes and long enough to stretch from temple to temple. Make thin horizontal slits for your eyes. Although these limit your field of vision, they will not frost up as sunglasses may do.

may prefer the classic gray worn and favored by mountain climbers since Sir Edmund Hilary wore them to the summit of Mount Everest in 1954, but a bright red union suit can have a certain cheering effect on a particularly cold day. Women's and children's underwear comes in a variety of prints and pastel colors.

Some camping and outdoor clothing suppliers carry winter underwear made of angora wool (blended with cotton and synthetic fibers), a luxuriously soft yarn that has the warmth advantages of lambswool without its scratchiness.

Net underwear has enjoyed a certain popularity among campers, but there is some disagreement about how significantly it contributes to keeping warm. Some feel that it works well when worn under knit underwear: the mesh hold a layer of warm air right next to the skin and at the same time permits easy evaporation of moisture. Others say that although this is true, it is relevant only for people who sweat profusely.

Down-filled and polyester-filled underwear are also available for wear in extremely cold climes. It is very heavy and bulky (despite manufacturers' claims to the contrary), and can inhibit movement. This is one case in which ventilating net underwear is a must, because it prevents excessive perspiration from reaching the filling, which would inhibit its insulating powers.

Women who wear skirts and dresses need not despair of a warm underlayer. Winter underwear is also available in briefer, less sports-oriented styles. T-shirts, camisoles, and knee- and shorts-length pants can be found in ribbed cotton and polyester knits as well as light wool blends. Your grandmother has known about these for years; there's no reason why you should not take advantage of the extra warming layer they provide without the bulk (and complete coverage) of thermal undies. The knickers or shorts are particularly useful as draft protection when you wear a skirt. Look for them in the lingerie department of most ladies' clothing stores.

A word about comfort. Winter underwear should not be too tight-fitting. In fact, if the waffled fabric of thermal underwear is stretched too much, it will lose some of its insulability. Cuffs, both on pants and skirts, should be elasticized to keep out drafts. If you choose two-piece underwear, the shirt should be long enough to tuck securely into the pants so it does not ride up every time you bend over.

That's the first layer, but underwear does not necessarily have to be worn *under* anything. If you like the look, there is no reason

---

Freezing

What It Is
Tissue destruction due to prolonged exposure to the cold. A true medical emergency.

Where It Strikes
Lower legs, feet, arms, hands.

Symptoms
Large blisters form, both on skin surface and on underlying tissue. Area feels cold, hard, and solid to the touch; will be completely unyielding when touched; joints will not move.
No sensation at all in affected area—victim may stumble along, not realizing that his foot is frozen.
If tissue destruction is severe, gangrene may set in. Professional medical treatment must be sought, and skin grafting may be necessary.

Treatment
1. Wrap affected area in warm blankets or clothing.
2. Bring victim to warm area; build a shelter if necessary.
3. Give hot liquids. Coffee is best; alcohol is to be avoided.
4. Remove any clothing that is frozen or that might restrict circulation to affected area.
5. Continue treatment as for frostbite. *Do not* permit victim to walk on a thawed foot. If one is alone and must walk to receive aid, no attempt should be made to thaw the feet until one has reached the destination, since refreezing presents a greater danger of permanent tissue damage.

Prevention
Follow the rules for frostbite prevention. Avoid getting excessively tired or hungry during prolonged periods outdoors.

why your brightly colored winter underwear cannot "come out" and be worn as a shirt, exercise suit, or pajamas.

Shirts, Sweaters, and Vests

It should come as no surpirse that cotton and wool are the best choices for the shirt, sweater, and pants layer. Perhaps the best place to start is with a cotton knit turtleneck shirt. They are always available at army-navy stores at modest cost; or you can buy a designer turtleneck at a boutique for quite a bit more. The important thing is that it be made of cotton (or cotton and a bit of synthetic for durability and easy care) so it will be lightweight and highly absorbent. Over the basic turtleneck, a cotton denim workshirt or other light cotton shirt that buttons down the front is often adequate for fairly warm indoor temperatures. If you have lowered the thermostat to reduce energy consumption, you might need a shirt made from a more substantial fabric. The pile of cotton corduroy shirts makes them warmer than a plain woven cotton shirt. Cotton flannel shirts are somewhat warmer; the napped surface acts as an effective insulator. They are naturally soft and slightly fluffy, and require no breaking in. Some cotton flannel shirts come with a double yoke, an extra layer of fabric across the shoulders to make things warmer yet. A cotton chamois cloth shirt is another good choice. Chamois cloth is a tightly woven doublefaced cotton flannel; it feels wonderfully soft against the skin because it is napped inside and out. It is also dense enough to protect against drafts inside and wind outside, and like other cotton fabrics that shrink slightly, its density increases with washing.

You can stay even warmer if you supplement cotton layering with a wool garment. Because of wool's greater bulk, one layer can take the place of two cotton ones. A lightweight, closely knit cashmere sweater is very warm and will look stylishly trim. Bulkier are Shetland and Icelandic sweaters; made from the wool of sheep raised in cold climates, they are natural insulators. One of the best 100 percent wool sweaters is made by the U.S. Navy, and although it only comes in one color, it is terrific for keeping you warm.

A vest, either wool or down-filled, will add an extra layer of warmth to the torso area while leaving arms unencumbered. Although down vests are generally considered to be outerwear, they are nice to have when it's particularly cold inside too. Down vests are generally cut extra long in back so that they protect from the neck right down to the kidneys. If you feel that the quilted style is too sporty for urban wear, there is a down vest available that reverses to an unquilted side that is quite businesslike. Business suits for both men and women often come with matching vests. And, of course, there are woven, crocheted, or knitted woolen vests, and cotton patchwork, denim, or corduroy vests in a variety of colors and patterns to spark up your outfit while affording you an extra layer of warmth.

The chest and back between the shoulderblades is another area of cold vulnerability—that's where your lungs reside—and it is often comforting to have an extra layer there, too. Dickeys are a good solution. They come in wool knits, and even filled with down and trimmed with a knit turtleneck. A scarf or shawl thrown over the shoulders and tied in front across the chest shouldn't be forgotten either. It can work as a fashion accessory as well as a chest protector.

Running Warm

Anybody can run during the warm months, and everybody seems to. But the joys of cold weather running! Few runners! Invigorating air! Shifting light and colors as nature changes gears!

Many novice runners shy away from running during the winter for fear of the cold. Curiously enough, however, the biggest problem for winter runners is the same as for summer runners: overheating. Track coaches and veteran marathoners alike assert that cold weather itself poses no particular problems for runners, which is why you'll see cross-country runners in shorts and T-shirts even during the coldest months.

Nevertheless, if cold weather running leaves you cold, the following are specific suggestions for keeping warm.

Clothing

Clothing that's either too tight (restricting the body's ability to let off heat) or too heavy forces the body to overheat.

Peter Schuder, track coach and director of physical fitness at Columbia University, recommends the "layered look": T-shirt, long-sleeved thermal undershirt, and light sweatshirt for the upper body; shorts, thermal long johns, and sweat pants for the legs; white cotton socks and running shoes; and most important, gloves and a cap or hat. He notes that 85 to 90 percent of one's body heat is lost through the top of the head. Bald? It's 95 percent.

Other clothing recommendations: pantyhose or tights (for women and men); coconut oil for the legs, if one layer is not enough warmth and two is too much; a ski mask for windy days; lip balm or Vaseline for the lips. Be careful, however, with oils applied to the face to "protect" your skin. They interfere with your body's temperature regulators.

Most runners will tell you they get cold only when they're overdressed and begin to sweat profusely. Moral: Strip down for comfort.

Warming Up

Coach Schuder recommends first a series of indoor static exercises, mostly stretching, followed by dynamic outdoor exercises. Begin with ten to fifteen minutes of exercises that stretch the calves, Achilles tendons, and hamstrings. Their purpose is to counteract the tendency of muscle and tendon sheathing to hypertrophy, i.e. remaining the same size while exercised muscles and tendons grow—painfully—inside the unexpanded sheathing.

Remember: Dress lightly for indoor exercises. Do them slowly and easily. Don't go outside sweating.

Follow with ten to fifteen minutes of outdoor exercises that "exaggerate" the running form: jumping jacks, running in place, short—30-yard (27-meter)—wind sprints, and skipping. These stretch the hamstrings and warm up the quadraceps (the four front muscles of the thigh). Coach Schuder also recommends random arm movements to keep the arms from tightening up.

Now you should be ready. And warm.

Eating

Some people can eat a full-course meal thirty minutes before running with no trouble; others can eat only after running. This is a personal matter and can be determined the first day you run.

If you do eat before running, you may find that you stay warmer if you drink hot tea or coffee rather than cold fruit or vegetable juices. Or you may not.

Crummy Weather

It's hard enough to get up in the dark at 6:00 AM in the middle of winter. When it's raining or sleeting, the temptation is to practice REMs instead.

But winter precipitation (for psychological uplift use this term instead of *rain, sleet*, or *snow*) is merely an inconvenience. A peaked hat over a ski cap keeps the elements off your face. A lightweight, roomy plastic poncho and an extra pair of socks will keep you dry. You'll get wetter from sweat.

Icy conditions can cause injuries: strains, twists, and bruises if you should fall. If conditions are intolerable and you're sliding around crazily, go home.

Breathing

When it's extremely cold it may be easier to breathe through a ski mask or scarf. There is, however, no truth to the horror stories about "frozen lungs" and pulmonary mishaps from breathing frigid air. By the time air reaches the back of your throat it has warmed to body temperature.

Some people suggest breathing through your nose. If you can do it, fine, but it's a real feat to get enough air to your lungs this way.

A Stitch in the Side

You're two miles from home, it's freezing cold, and suddenly you feel that familiar pain in the side. The stitch is a common running ailment, yet there is no agreement on what causes it. It does seem to occur less frequently the more you run. Nor is there universal agreement on how to get rid of a stitch, although exercising it out is preferred to indolence. Keep moving, even at a walk. Breathe more deeply using your stomach muscles. Push against a tree or wall. Do exercises that stretch your stomach.

Stopping

Wind down slowly. Walk the last 200 yards (187 meters) to your door or car. You can get as sick being overheated inside as you can from standing outside sweaty.

If you drive, a blanket to throw over your shoulders is an excellent idea. Otherwise, you can get a bad case of the shakes sitting immobile in a cold car.

For the perfect conclusion to a midwinter run, gather about a hundred friends, whip off your running cap, and watch them ooh and ahh as your head steams like a cheap cigar. This is a good sign.

—Stuart Fischer

Pants and Jeans

When it comes to layering the lower half of the body, similar principles apply, but there are also special considerations to keep in mind. In a sense, your legs need less protection than your torso; there are no vital organs located in the legs, and if you are even the least bit active, circulating blood in your legs does a good job of keeping them warm. This is not to say you can run around unclothed from the waist down, but it does mean you can dress your lower body less bulkily, permitting greater comfort and ease of movement in your legs, which, after all, are meant to move. For this reason, choose trousers that provide adequate warmth and insulation without a great deal of bulk. For indoor wear, you will probably find cotton denim or corduroy jeans suitable; if your thermostat is on the low side, add tights or long underwear. For colder indoor temperatures and for outdoor wear, wool trousers are best. Army surplus pants and thirteen-button sailor pants (though these take forever to get on and off) will last a long time and provide lots of warmth. Pants made from woolen whipcord are not only warm but virtually windproof. Wool blended with 15-20 percent nylon will be nearly as warm as and much more durable than 100 percent wool; if you do not find wool-blend pants on the racks of your favorite clothing store, try camping and outdoor sporting goods stores, as well as mail order outlets. Look there also for field pants made of cotton and polyester treated to repel water and lined with nylon/wool blended fabric to provide warmth and insulation. If your skin is sensitive, look for wool pants with lining unless you plan to wear long underwear as well. And remember that wool content is a less reliable indicator of a garment's warmth than is weight, density of weave, and type of finish.

If it is terribly cold or windy and you need to spend a considerable amount of time outside, nylon taffeta wind-pants and down- or fiber-filled outerpants are worth looking into. These zip or pull over a lighter pair of pants and provide an effective barrier against the worst that winter has to offer.

Mountain climbers and cross-country skiers like knickers for sporting activities, but they can be adapted to everyday wear. Look for them in strong wool tweed or cotton corduroy, and be sure to add a thick, colorful pair of long socks to cover your legs where the knickers don't.

Outerwear

When it comes to the outermost layer, a number of different factors will determine how you choose to dress. Taste and lifestyle, age and budget, and the sort of climate you'll be spending your time in all come into play.

The winter coat or jacket made of wool or wool blends, lined or not, is the outer garment of choice for most urban dwellers. Differences in style are myriad, of course, but certain structural features are worth looking for.

If you will have only one outer garment for the winter, make it a coat that covers you below the waist; you need the warmth there because your kidneys are particularly vulnerable to the cold and short jackets do not provide the protection needed. And make sure the coat buttons well past the waist; it is surprising how frequently a design idea will win out over good sense, leaving you with the lower half of your coat flapping open in the wind. Check the back vents as well. The best thing is to have none; second best is a pleated or gusseted vent so you will be able to take long strides without exposing the back of your legs to the cold. Avoid coats with skimpy collars, and certainly those with no collars at all. Turning your collar up against the wind can be an effective second line of defense.

Although they are quite fashionable, capes are not the best cold weather garments. Their loose fit and armhole instead of sleeve do not provide adequate protection against wind and weather.

Storm coats, usually made of water-repellent poplin and lined with pile or wool, are a good choice if your climate is on the wet side. If you encounter more snow and subfreezing temperatures, water repellence is less important than insulation, and a lined wool coat is your best bet.

If you like a sportier look or engage in strenuous activities for which a coat would be too confining, sandwiching warm layers under a windproof/water resistant shell is a good idea. You can affect this with as pedestrian a garment as a rain jacket or coat large enough to accommodate several heavy sweaters and still permit you to move. Or you can become more sophisticated, and spend more, with anoraks, windbreakers, or rain shells made of nylon (coated with neoprene if you want water repellence), cotton and Dacron, or treated cotton canvas. Again, thermal underwear, wool shirts, and heavy sweaters get layered underneath and topped off with the outer shell. This is an arrangement for which a down-filled vest makes sense too. Look for one with kidney-flaps or extra long back.

When you are going to be wearing a lot of layers anyway, you may want to consider a down or other filled garment. These have become extremely popular in recent years, and now you can get anything from a light jacket to the so-called expedition parkas that will keep you warm at 50° below zero.

Before you spring for one, though, you should consider the climatic conditions where you live. A down parka is terrific if it stays cold once it gets cold and most of your precipitation is snow. But if

Recycling for Warmth

There are many recycling projects that make use of those moth-riddled hand-knits or the old college sweaters so steeped with memories that you can't bear to discard them. The sleeves can be made into leg warmers for children (see page 70) or sewn inside the sleeves of a not-very-warm coat or attached over the sleeves of another sweater.

The body of any sweater can be cut apart and trimmed with braid or crochet to make an unusual vest. If you have any garments with suede-patch elbows, use those on the knees of your children's clothes.

Four-Armed Sweater
1. Cut off the sleeves of the larger sweater at the shoulder seams. Zigzag the edges.
2. Slip the larger sleeves over the smaller ones and handstitch in place, turning the edges under about ½" (1.3 cm). Fold up the bottom edge of the larger sleeves and hem to show a good portion of the sleeves underneath.

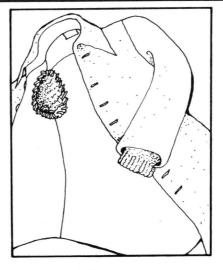

Doubled Sleeves for Coat or Jacket
1. Use a sweater with ribbed cuffs because this helps seal off the drafts that creep up your arm. Cut the sleeves off at the shoulder seams. Machine stitch the edges to prevent ravelling.
2. Place the sweater sleeves inside the coat sleeves. Turn under the inner sleeve edge about ½" (1.3 cm) and baste together the shoulder seams by hand.

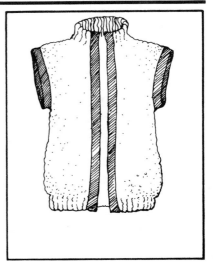

Vested Interest
1. Run a double zigzag stitch down the front center of the sweater to prevent bunching or ravelling while cutting.
2. Cut straight down the middle. Topstitch some contrasting braid on all the edges. Buttons and buttonholes could also be added.

—J.M.

winter temperatures in your area tend to hover around freezing and much of your precipitation is a chilly rain, your parka won't provide you with the protection against water that you need, and too often it will end up a lumpy, soggy mess, with you shivering inside it.

But if you have a warm rain or storm coat for wet days, and still want a parka for the blustery, windy ones, you will be amazed at the variety and versatility of the garments you can buy. Here are some facts to help you figure out which one is right for you.

Down-Filled Parkas and Synthetic Alternatives

Shell Fabrics

The shell of filled garments can be made of various fabrics. Nylon is by far the lightest of the fabrics used. It is also the thinnest. This makes it preferable for backpacking, mountain climbing, or other situations in which space is at a premium.

In exchange, nylon is the most delicate of the fabrics. Even though a heavier-gauge filament can be used in the weave to make it more resistant to abrasion, nylon is not really suitable for heavy outdoor work. Its proneness to tearing when punctured also makes it a risky fabric to depend on for small children who like to roll around and bang into things.

At the same time, nylon can be woven more tightly than other fabrics, which puts it at the top for its windbreaking ability. This can be a real plus for skiing, hiking, and other activities that can expose you to the winter wind.

Nylon for shells generally comes in two weaves; a smooth one, sometimes called taffeta, and one called rip-stop. Taffeta is the one familiar to us in most windbreakers, and it comes in various weights. The heaviest of these begins to compete with lighter alternatives for resistance to abrasion, but it is still subject to tearing. Rip-stop, a fabric that has emerged from the technology of camping, is an extremely light and close-woven taffeta into which much thicker threads have been incorporated to divide the fabric into boxes about 1/8" (.3 cm) on a side. When a tear reaches one of those thicker threads, it is stopped.

Cotton is the most durable and, when treated, the most water repellent of the outer shell fabrics. It is also the heaviest and the most expensive to use. It is, for that reason, unusual to find pure cotton used in outerwear. Generally it is blended with polyester, in some proportion, to make use of the properties of both.

Treated cotton will need care in laundering. By law, the label must tell what the treatment has been, and manufacturers generally provide laundering information.

The best cotton blends tend to be in a proportion around 60 percent cotton, 40 percent polyester. This combines the durability and waterproof qualities of cotton with the lighter weight, closer weave, and wrinkle-free virtues of polyester. It is less windproof than nylon, but still better than cotton alone. This is the best fabric for working or being very active in cold weather.

Another popular blend, which can cause some confusion, reverses the fiber proportions to 60 percent polyester, 40 percent cotton. Needless to say, this is a cheaper, less desirable fabric than the 60:40 mentioned above. It still retains some of the virtues of cotton, but it is not that much more rugged or water resistant than the thicker grades of nylon.

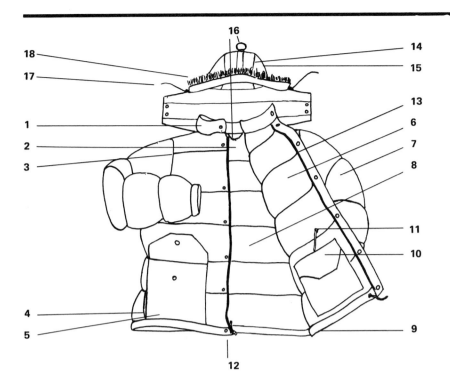

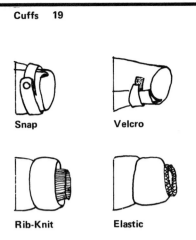

Snap   Velcro

Rib-Knit   Elastic

19   Cuff should be tight against your wrist
to keep heat in, loose when you want
ventilation. Some cuffs are plain elastic,
whereas some have a recessed rib-knit
inner sleeve. Some cuffs have a system
for tightening and loosening them with
snaps or velcro.

Cargo Pocket   20

Velcro   Bar-Tacking

Snap

Double-Sewn   Bellows Pocket

20   Cargo pockets should have flaps that
give free access. Many close with velcro, which breaks less easily than a
snap and will not get clogged with ice
or snow. Some, called bellows pockets, expand for more room and less
stress. Whichever you choose, your
pockets should be easy to handle with
your mittens on and big enough to hold
your mittens when you take them off.
—Alan Ravage

**Twenty things to think about when you buy
a parka**
You have been dreaming about your ideal
parka, but now you're in the store, surrounded by real ones, and several of them
look good. The twenty items on this page
are capsule reminders of the things that
make one parka different from another and
some parkas better than others. They will
help you make that final choice of the one
that is really best for you.

1   Collar should be filled and come at
2   Fit should be loose enough to leave air
trapped between you and the parka
lining. It should also be ample enough
to let you wear sweaters and other
layers comfortably.
3   Some people prefer set-in sleeves, some
raglan style. Parkas come both ways.
4   Pockets for warming your hands are a
good idea for when you take off your
mittens or gloves. But they have to be
insulated to be effective.
5   Frequently the hand-warming pockets
are set behind the "cargo" pockets.
But watch out for a pocket that just
has two openings. You may think you
have put your wet mittens into their
own compartment, but when you slip
your hands into the other opening,
there are the mittens, damp and
freezing cold.
6   Down should be evenly distributed in
the horizontal chambers. Vertical quilting is there for looks only, and may
even cause cold spots. It does nothing
for the down. Feathers showing
through the seams means that the
parka was sewn after it was stuffed,
and the filling may be very uneven.
7   Check to see that the filling is thick at
the armpits. It may seem less comfortable that way, but it is one of the most
important places to keep warm and a
place where a lot of parkas are skimpy.
8   Lining should be free of holes or snags
or tears. Check that there are no loose
threads or compartments where the

stitches have pulled out. Turn the
sleeves inside out and check all the
way down to the cuff. When you wear
your parka, be careful of rings, watches,
and other sharp things that can tear the
lining when you put it on.
9   The bottom should reach your thigh
for maximum warmth.
10   Many parkas have an inner pocket for
valuables. Be sure it closes securely.
11   Good parkas have drawstrings or cinch-
cords inside to keep warm air from
escaping out the bottom. Belts do the
same thing but, because they crush the
filling, they provide less warmth at that
spot. Whatever the system, it should be
at your waist, not above or below it.
12   Test the zipper before you buy a parka.
It should be in good condition and
work smoothly. Larger zippers have
less tendency to snag and are easier to
handle with mittens on. Plastic zippers
tend to freeze less than metal ones and
some are even "self-lubricating." Two-
way zippers that open from top to
bottom are useful for ventilation.
13   Zipper should be protected by an insulated flap that snaps down to prevent
serious heat loss.
14   Hood should be filled too.
15   Detached hoods meet a greater variety
of conditions, but attached ones are
warmer.
16   Some parkas with attached hoods have
a hanger ring on the hood. Ones with
hoods that come off have a loop or
chain near the collar. Whichever system
your parka uses, it should be strong
and attached firmly. Brush off your
parka when you hang it up so that
melting snow does not dampen the
shell and filling.
17   Hood should tighten about your forehead and temples to retain maximum
heat. Most do this with a drawstring.
18   Fur trim will not keep your face warm.
It should be detachable to make washing the parka easier.

Many outer garments use nylon in their construction and cover it with an extra shell of cotton blend. In this case, a lighter grade of nylon will be adequate.

A recent medical discovery has led to a new development in fabric for parkas and other outerwear. Called Gore-tex, this material was developed as a synthetic replacement for arteries in heart surgery. Related to Teflon, it is a film that contains nine billion pores per square inch (6.5 sq. cm). This means that it can "breathe"—let air in or out—but that its pores are too small to let water pass through. It is generally laminated between layers of nylon to produce an extremely lightweight, virtually waterproof material. Although it is still expensive, most manufacturers are offering at least one line of garments made with it. Aside from its price, its other drawback is that it is just as delicate as nylon and, although it can be patched, should not be used for rugged work or play.

How To Buy A Parka

You can buy a parka in just about any clothing store in the country now, but you still have to shop carefully to get one that is just right for you. If you get the wrong parka, you may find that you have shelled out good money for a garment that is more trouble than it is worth.

The first thing to think about is what a parka is and what it does. Simply stated, a parka is an outer garment consisting of an outer shell and an inner lining with an air-trapping filler between them. The more protection you need, the thicker the layer of filling has to be and the more thoroughly the shell has to be isolated from the lining. Those are the principles, whatever the claims on the label may be. Some manufacturers call their heaviest parkas "expedition" parkas, and strive for protection against extreme cold for long periods of time. That doesn't mean that they are more rugged or better made; it means that you will roast inside it unless you spend most of your time outdoors under conditions of severe cold.

Label names can go from the technically specific to the vague or suggestive to the outright deceptive. Some manufacturers will call their parkas jackets, some will call only their longer jackets parkas. In

fact, the word *parka* is Eskimo for jacket. But be careful when you see the words *ski* or *backpacker parka*—the catch is that those are activities that heat you up and require more agility, so the filling material will be thinner than that of other parkas. Don't buy a parka for the name on the label; buy it because it will do what you need it to do for you.

The first thing to think about is how warm you need to be. Think about what you will be doing when you are wearing your parka, how cold it is likely to be outside, and how long you are likely to stay out before you can come inside to get warm. You will probably find that you need a parka that can be regulated to retain more heat or to provide ventilation. Depending on the garment, you can do these things by adjusting the cuffs, hood, collar, front zipper, and drawstring. Each of these parts of a parka corresponds to a point on your body that generates heat—your skull, neck, wrist, and vital organs. By closing them within your parka you retain their heat. By exposing them to the cold, you dissipate it. To be worth the money you pay for it, a parka should be able to protect you at every one of these points, unless you yourself decide on an alternative protection (enough people prefer hats to hoods that many manufacturers sell their hoods separately as an option). But your parka should have enough ventilating mechanisms to meet a variety of conditions.

The next thing to think about is convenience; don't buy just for looks. Think again about what you will be doing when you are wearing your parka, and get one that suits those needs. If you will be working or playing roughly in it, get one with a rugged outer shell, maybe even one that is quilted from the inside so there are no seams showing to snag or wear. But if you are going to be flying to a ski resort, with a sports and evening wardrobe, you may want a rip-stop nylon one that can be compressed to nearly nothing. Don't forget that you will be wearing mittens or gloves, check to see that the zippers, flaps, and snaps are easy enough to manage while you're wearing gloves. And remember that you may want to take off your hand coverings. Most parkas have hand-warming pockets, but they won't be any good unless they are insulated. Some parkas stick the pockets behind larger square pockets. This is a good idea, but these so-called cargo pockets still have to be insulated, Cargo

pockets are very convenient for mittens or gloves, tools, and other things you may want or need outdoors, from an axe to Chapstick. Make sure they are big enough and conveniently accesible. If you carry a wallet or other valuables, you may want to get a parka with an inside pocket. Make sure it closes tightly. A detachable hood is more convenient than one that just slides back.

Make sure your parka is well made. The way to test loft is to crush the filling and watch it regain the shape it had before. It should not feel limp or soggy or lumpy. If you are buying a down garment, make sure that the compartments are full and evenly distributed and that when you squeeze them you cannot feel any quills. The down should not come through the stitching; no matter that the salesman tells you that it is an indication of how well stuffed the garment is, what it really means is that the compartments were sewn after the garment was filled and when you have pulled the down away from the stitches there may be cold spots. If you really like that particular garment anyway, be sure before you buy it that there are no thin places in the filling. Zippers and snaps or Velcro should be heavy-duty enough to withstand the stress you and the cold weather will subject them to. Seams should be double-stitched or bar-tacked at points of stress, especially at the collar, the ends of the zipper, the pocket openings, and the flaps.

The feel of a parka may seem strange when you first put it on. It takes some getting used to. The important thing is that it should fit you comfortably under the outdoor conditions it was meant to meet. It should be big enough so that when you are wearing layers under it—sweaters, vests and jackets—there is air between those layers and the parka. At the same time, it should not slide all over you. Sleeves should have a rib-knit recessed cuff, or an elastic one, or some system for tightening it at the wrist with snaps or Velcro. It should have an inside drawstring at your waist, not above or below it. The collar should be snug without being tight. The hood should be big enough to fit you without compressing the filling, and it should close in a way that brings it tight around your face next to your eyes.

When you have found the style you like, try on several in your size. Because parkas are bulky, there are often minute variations from one to another, and one may fit you just that much better.

—Alan Ravage

Fillings

Fillings are either down or a replacement made of synthetic fiber. Each has its virtues, but the main consideration is the same: loft, how well the material traps and holds air in its structure. Although there is no hard and fast ratio of loft to warmth (and warmth is such a subjective matter), generally speaking, the thicker the fill, the warmer the garment.

Down is the insulating material with the lightest weight and the greatest loft. The best grades have a loft of up to 750, although most garments promise a loft of around 550 or 600. (The numbers refer to how many cubic inches one ounce of down will fill.) Even so, those in the know caution that the recent popularity of down had driven the cost up and the quality down, so that regardless of what the label says, you may still be getting a loft closer to 450. And that begins to be on a level with synthetic fillers.

Manufacturers in the United States often label their filler "Prime Northern Goose Down." What that means in that their down comes from sources in this country. The distinction they are making is between their down and down from Taiwan, Korea, or Hong Kong. There have been complaints that down from the Far East has been extended with duck and chicken feathers, which, besides having an inferior loft, tend to poke holes in the fabric.

There are two significant disadvantages to down. It is not waterproof, and when it gets wet it clumps and loses all of its insulating ability. This means that in seasons and climates where the temperature is variable and snow can become rain without warning, down garments can be less suitable than synthetics. The other disadvantage is that because down contains natural oils, it needs special attention when you are laundering it. By law, manufacturers are now required to put washing instructions on their products; they should be followed carefully.

Synthetic replacements for down are materials made from a continuous filament of plastic woven in such a way as to produce loft. Each manufacturer has a different name—the most widely used is made by Fortrel and called Polar Guard—but they are all basically the same. Their loft is about 400, less than good down. This does not mean that garments made with it are less warm; it just means that they have to be heavier and bulkier to get the same degree of warmth as a comparable down garment.

The advantages of synthetics over down are that they are substantially cheaper, easier to maintain, and will not retain water. This means that you can fall into the lake, wring out your parka, and wear it home; it will be clammy and damp, but the filler will insulate you against the cold. Down won't. You can throw a synthetic into the washing machine with the rest of your gear, tumble it dry, and it will be ready to wear again. Down is washable too, but it requires special treatment.

Since it's a continuous filament, synthetic filling does not need to be compartmentalized. This means that although garments made with it may be stitched with the characteristic quilted squares, they do not need to be. The filling will keep its shape without it, unless you tear it or pull it apart.

Head Coverings

When your mother admonished: "You can't feel warm if your

head isn't warm," she knew what she was talking about. The heat loss from a warm body via an uncovered head is considerable, great enough to make you feel cold all over no matter how swathed in down the rest of your body may be. So wear a hat or a hood, and make sure your ears are covered too. The more area above the neck that is covered, the warmer you'll be.

Down-filled jackets and parkas often have down-filled hoods, and those that can be pulled tightly around the face provide even more protection. A nylon shell hood provides good wind protection but not much insulation. Knitted ski caps or navy and coast guard watch caps can be coupled with the shell or worn alone. Or wear a beret, a stocking cap, a fur bonnet, or wrap a warm scarf around your head. Follow your own taste, but make sure the hat you choose is wool, rather than acrylic, for maximum warmth, and that it will stay firmly on your head no matter how hard the wind blows.

Some extra-warm choices include a balaclava hat; it can be worn rolled up cap-style, or rolled down to cover ears, chin, neck and forehead, leaving only a patch of face uncovered. If your coat or jacket does not come with an attached hood, consider a combination hood and dickey; close fitting around the head, it extends downward to protect neck and chest and upper back. Another sensible style is a hat equipped with storm flaps that fold down over ears and forehead. These are available in utilitarian versions or, for a bit of whimsy, as a Sherlock Holmesian deerstalker cap, or lined with pile or quilted fabrics (or warmer and more luxurious still, with shearling or mouton fur).

If wearing a hat is just not your style, invest in a knitted ear band or fur earmuffs. The large skin area of the ears contributes significantly to heat loss, so it makes sense to use ear coverings in conjunction with a hat too.

And don't forget your neck. A scarf is a must to keep your neck warm and slow down heat radiation. Added to that, fur collars not only look rather special, they can provide a great deal of warmth if they are turned up against the neck.

If your face really suffers in the cold, consider down-filled, doeskin, or woolen face masks. An air-warming mask, which fits over the mouth, nose and cheekbones, raises the temperature of inhaled air to an average of 18° C (65° F). If breathing in cold weather is agony for you, or if you have respiratory problems, this might be the solution. These masks can give a pretty weird look. Less extreme, if somewhat less effective, is wrapping a scarf over cheeks, chin, and

Chapped Skin and Chapped Lips

Rough, red, irritated skin is a common winter malady. Dry air indoors and strong winds outdoors rob skin of its natural moisture. Prevention of chapped skin involves a few simple tricks, and a "cure" is as close as your neighborhood drug store.
1. Wash your face and hands in warm, not hot water, and use less soap in the winter than you do other seasons.
2. Rinse face and hands with cold water and *dry thoroughly* with a soft towel.
3. Use a mild lotion or cream daily to restore moisture and protect the natural moisture of the skin.
4. Add bath oil to your bath to protect your whole body from moisture loss.
5. Avoid licking your lips or breathing through your mouth. If you must

breath through your mouth, apply petroleum jelly or a commercial product such as *Chapstick, Blistex* or *Bio-Stik* frequently.
6. If skin does become chapped, do not wash with soap and water for several days. Instead, use a cleaning cream or oil or other soap substitute.
7. Apply cold cream, vanishing cream, or another emollient lightly before going to bed. Or make your own chapped-skin balm by mixing equal parts of lemon juice and glycerine, which is safe for both skin and lips. For chapped hands, try covering your hands with castor oil, petroleum jelly, or any hand lotion containing lanolin, vegetable oils and fats, put on a pair of white cotton gloves, and enjoy an overnight treatment.
8. If the air in your house is excessively dry, consider using a humidifier.
9. In very rare situations, chapping of skin,

and especially lips, may indicate a vitamin deficiency or other medical problem. If your chapped area does not clear up within a reasonable amount of time, see a doctor.

mouth. Your breath will create a warm atmosphere and the scarf will trap it near your face.

Hand Protection

Your hands should be diligently protected outdoors. They are generally the first part of your body to react to cold temperatures, and once they get cold, it is hard to get them warm again.

Mittens are the best hand covering for very cold temperatures because keeping all the fingers in the same compartment permits them to warm each other; with gloves each finger is on its own. Wool mittens will keep your hands warm; even better are down-filled or nylon-covered polyurethane foam mittens which are said to be warm in temperatures as low as –17° C (0° F). There are many other styles and fabric possibilities, of course, including woolen inner mittens with water-repellent and wind-resistant outer shells made of leather or nylon. Fur or fur-lined mittens are luxuriously warm; shearling mittens combine the softness of fur with the strength and durability of sheepskin suede.

If you need to use your fingers outside and mittens seem too awkward, there are two solutions. One is hunting mitts, down-filled mittens with a flap-protected slit that allows you to slip your fingers out to strike a match, take a picture, or anything else without removing the whole mitten. Another possibility is to wear nylon, silk, or wool glove liners with any type of mitten. If you must remove your mittens the liner will protect your hands for the brief time they are exposed to the air.

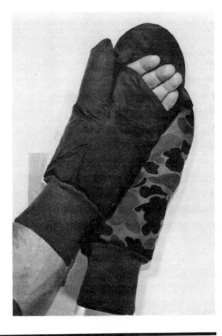

Gloves do a decent job of keeping your hands warm when temperatures are moderate; leather gloves lined with wool, fur, pile, or shearling will provide enough protection if you need to have serviceable fingers outdoors.

Make sure your mittens or gloves are long enough to overlap the sleeves of your outer garment. Wind and snow are attracted to naked wrists like a magnet. Be sure to close that gap! Blowing into your mittens or gloves before you put them on will warm the insides for quite a bit of time. If your hands get unbearably cold, a quick way to warm them is to tuck them under your armpits.

And whether you choose gloves or mittens, take care that your hand coverings stay dry. Or have a standby pair or two in case the

Heat-producing Extras

If your hands and feet are particularly sensitive to cold, or if you plan to spend a long period of time outside and are wary of frostbite, several mechanical aids are available. Depending on your appetite for gadgetry, some may appeal more than others.

Comfort Products of Aspen, Colorado, has developed a pair of gauntlet-length electric gloves that incorporate a rechargable power unit operable by a single switch at the wrist. The gloves are lined with reflective material. The same company offers electric insert soles with a rechargeable belt pack; they come in two sizes, one for regular wear, one for ski boots. Electric socks for boots, which operate on a D cell battery, are made by the Timely Products Company.

You can also find auxiliary warming devices that are not worn, but give quick and welcome heat in emergencies, or when you are sitting still for long periods of time (e.g., spectator sports, ice fishing, duck hunting). *Instant Heat* is a chemical-filled plastic pouch that gives off heat from a chemical reaction that takes place when the pouch is squeezed. The bag can be applied to any part of the body that needs instant heat. Chemical heat can also be supplied by the reusable *Scotty Chemical Handwarmer,* which operates when snow or water is added to the contents of a pouch. The heat will last up to six hours. *Hotstik* is a solid charcoal-like substance covered by an asbestos pouch. The stick burns for four to eight hours; the asbestos acts as a barrier so that the product provides warmth, but not burning heat.

By far the most popular device, used by campers and mountaineers, is the catalytic handwarmer. Burning lighter fluid, catalytic fuel, or white gasoline, products like the *Optimus* and *Jon-E* handwarmers provide heat for twenty-four hours on one filling. You can regulate the amount of heat produced by the Optimus model.

one you have been wearing become wet. Nothing feels colder than wet hands, with the possible exception of wet feet.

Foot Coverings

Feet, like hands, are early victims of cold. Because they are the part of the body most distant from the heat center, they are the last to get warming blood and the first to notice when the supply is curtailed to provide more warmth for the torso area. That is why one way of keeping your feet warm is to wear extra clothing on your midsection. If you become overheated, your body will correct that by sending the over-warm blood to your extremities. And if head and hands are adequately covered, your feet can be the major beneficiary.

Dressing your feet for warmth and comfort begins with choosing the right socks. Wool socks are the best insulators and absorbers of foot perspiration. Two pairs of wool socks are recommended for winter hiking or long periods outside. You can wear liner socks made of cotton, silk, or synthetics if the feel of wool bothers you. Some liner socks are made from synthetic blends designed to wick off moisture; others are lined with terrycloth. Tennis socks come with terry lining and they might be the perfect inner pair for winter wear.

The longer the sock, the more area it will keep warm. For skirts and knickers, long socks are a must. Colorful knee socks or patterned

---

Legwarmers

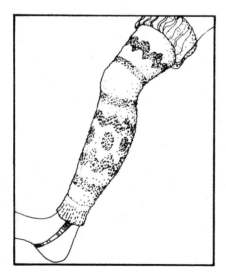

Long used by dancers to prevent muscle cramping during strenuous workouts, knitted legwarmers have recently become the fashion rage. But even if your passion is not for the latest chic, legwarmers have a

great deal to offer. They provide an extra layer of knitted insulation from ankle to upper thigh, and because they are basically just a long tube, you can adjust the area they cover to suit your own comfort—roll the tops down to the knees or the bottoms up past the ankle to the calf, or pull them all the way up and all the way down to cover your whole leg.

Legwarmers are available in lots of different colors and patterns. They are made

from synthetic knits, blends, or 100 percent wool, and that, as well as where you get them, will affect the price. You can pay as little as $4, as much as $30 or even more. Or you can make your own; next to a scarf, legwarmers are probably the simplest knitting project imaginable.

Materials
One pair each of no. 3 and no. 5 knitting needles, or whatever size you need to achieve gauge.
Three 2-oz (56.7 g) skeins of two-ply sport yarn
Optional: 2-3 yds (1.8-2.7 m) 3/4" (1.9 cm) elastic
Gauge: 6 stitches per inch; 7 rows per inch (15 stitches per cm; 18 rows per cm)

Measurements: Before you begin, measure ankle to upper thigh, and measure around ankle, around leg just above the knee, and around upper thigh.

Legwarmers you buy are usually one-size-fits-all. This pattern permits you to make your own to fit more exactly, but remember that there is a lot of lateral stretch in a knitted tube, so measure the length carefully, but class yourself widthwise as small, medium, or large. The instructions are for medium; small and large appear in parentheses in that order whenever there is a difference in the instructions.
With no. 3 needles, cast on 54 stitches. (Cast on 60 if you plan to wear the legwarmers over your boots rather than tucking them inside). K1, p1 in ribbing for 3" (7.5 cm), then change to no. 5 needles and increase 1 stitch at each side every sixth row 21 (20, 22) times; then increase 1 stitch at each side every fourth row 10 (8, 11) times. Work on 116 (110, 120—plus 6 if you are knitting for outside-boots style) stitches until piece measures 4" (10 cm) less than your ankle-to-thigh measurement.
Change to no. 3 needles, and decrease 1 stitch each size by knitting together. K1,

p1 in ribbing for 4" (10 cm). Bind off. Repeat for other leg.
Finish by sewing back seams together. Block carefully so both legs are same length and width.

Variations: Instead of a ribbed top or bottom, knit entire piece in stockinette stitch but add 1" (2.5 cm) at each end to overall length. Turn over extra inch and sew a casing. Pull through a 3/4"- (1.9 cm-) wide length of elastic to fit ankle and/or thigh.

Elastic stirrups at the bottom are handy if you plan to wear the legwarmers inside boots or shoes. Attach one end by sewing it to bottom edge; use snaps to attach other end.

Choose solid color or make stripes of any width by changing colors according to your own design.

—B.K.

legwarmers, which stretch from the ankle to the upper thigh, are cheery on a wintery day. Both can be worn over tights. Legwarmers and extra-long natural wool socks can even be worn over tight-legged jeans. To be avoided are socks with individual toe compartments. Charmingly whimsical as they are, they have the same disadvantages as gloves.

Conventional socks, like mittens, ensure a warming together-ness. Basketball socks have wide stripes above the ankles and stretch to midcalf, and you might find a pair of brightly striped wool soccer socks that will extend over the knees of most women. Hunting, fishing, and skiing departments stock more conservative wool and wool-blend socks, but these promise even greater warmth than other sport socks. Buy good quality (new or virgin) wool socks because the cheaper kinds (reprocessed or reused wool) are more likely to mat and lose their insulating ability. If wool socks have nylon-reinforced toes and heels, they will last longer.

No pair of socks, no matter how warm, will keep your feet comfortable if they are damp from sweat, or are too large (wrinkles equals blisters, it's as simple as that) or too small (we are aiming for trapped air, not trapped toes). So pay close attention to the con-dition as well as the composition of any winter socks you wear.

Boots, the fashionable and sensible winter footwear, come in a wide variety of styles, and your choice will depend on how and where you intend to use them. Even urban-style boots are usually better for winter walking than regular shoes. Many come with pile or fur linings and, with properly chosen socks, will keep your feet warm and dry. Hiking and camping footwear, though bulkier and more heavyweight than casual boots and shoes, is enjoying popularity even in the city. Regardless of what you wear on your feet, there are certain basic principles to keep in mind.

Buy boots that are roomy enough to take heavy socks and still provide your toes with wiggling room. The air in there with your feet will be warm and insulate well against the cold. Even if you have the best boots in the world stuffed with the warmest socks imaginable your toes will still feel like icicles if they are cramped together or pushing up against the front of the boot. So when you go boot shop-ping, wear the weight and thickness of socks you plan to wear all winter long.

The ideal boot should keep a good part of your lower leg warm, should be made from materials with good insulating properties, and should have pliable uppers. Since much winter walking involves nego-tiating snow banks, puddles of slush, and the like, your boots should also keep your feet dry. In this respect, a moccasin-toed boot is not the best idea; the stitching will catch snow and hold it until it melts onto your foot. In general, look for boots with little or no stitching at ground level, since stitched areas have large potential for leakage. Beware also of boots that zip and tie. A webbed tongue or zipper placket is the only reliable insurance against leakage, so make that a primary feature to look for.

Smooth leather boots can be treated with a silicone spray to make them moderately water repellent without affecting their breatheability; a heavier treatment of silicone wax will make them waterproof, but then they won't breathe. Mink oil, sometimes mixed with silicone in a pastelike compound, will also water condition smooth leather.

If the weather you are most likely to encounter is wetter than not, you might prefer a boot with leather uppers bonded to rubber bottom and soles. Sorel makes such a boot, as does L. L. Bean, in various styles. Some come with felt inner boots, which are removable for drying and help insulate an otherwise unlined boot.

Unlined rubber boots are not a good idea, nor are unlined leather-look boots made of polyurethane. These are essentially warm weather rain boots and will not insulate your feet against the cold. If you need the impermeability of rubber, be sure to look for boots with plush, felt, or foam lining. (The thin flannel often found in rain boots does not count as lining. That's just to help your foot slip in.)

Heavy-duty winter hiking boots may prove too awkward for normal activities. The uppers are quite stiff and the soles are rigid so they won't flex when you walk. They are, however, very warm and very leakproof. They often come with foam insulation around the foot and ankle. Some are completely foam lined; others have felt inner boots. One of the great advantages of hiking boots in their cleated or lugged soles, a significant safety feature when it's icy underfoot. Of course, you can buy cleats that strap onto any boot or shoe to help you maneuver on ice.

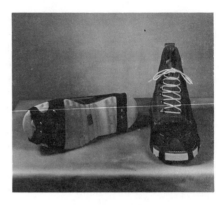

An insulating inner sole is something else that can be added to any boot or shoe. (Be aware, though, that they take up room, so you may have to wear a shoe a half-size larger to ensure that your toes still have wiggling space.) Inner soles are available in felt, foam, nylon mesh, lambswool, and even an aluminum Mylar-backed nylon. L. L. Bean has a cold-proof inner sole with a built-in arch that is made of sheepskin stitched to a leather sole. Felt has a tendency to blot up foot moisture, so the liners have to be removed after each wearing to dry out; nylon mesh is a better bet if your feet sweat a lot. The lambswool provides a warm cushion; and the Mylar works by blocking the outward radiation of foot heat (you wear them shiny side down).

Tucking your pants into boots will cut down on drafts and snow leakage; and if you put long socks over your pants legs before you put on your boots, the pants will not ride up as much. In very cold, windy, or wet weather, a set of gaiters in canvas or nylon are handy. These slip over your boots or shoes to close the gap between foot and pant leg.

Take good care of leather boots. They are usually a large investment, but if properly cared for they can last for years. When you come in after a snowy, wet day, you will probably notice salt stains on the boots. Sponge them off thoroughly with clear water, then stuff the boots with newspaper and stand them up to dry. Keep them far away from radiators, hot-air registers, fireplaces, and any other source of drying heat or the leather will dry out and crack in no time. Leather boots should, in any event, be treated periodically with a good shoe wax or cream to keep the leather supple. This is not the same as silicone-treating them, and one cannot be substituted for the other.

And once you're home with your boots drying in the corner, you can slip into a pair of slipper socks with leather soles to keep your feet warm even if the floor is not. Warmer still are down-filled "socks" or booties, some with leather or leatherlike synthetic soles, some without. Both are good choices for foot covering around the house, and the soleless "socks" are welcome companions in bed when it's really cold. And then there are the endless styles of slippers, lined with pile, lambswool, flannel, and more.

Keeping Kids Warm

Notions about children and cold weather have changed a great deal since the times when infants were swaddled and the common cold was so named because it was erroneously believed to have some connection with exposure to chill temperatures. Keeping children warm involves most of the same principles as keeping adults warm, with these exceptions: Children are generally more active, indoors and out, and have more efficient circulatory systems; their healthy young blood vessels have shorter distances to travel than the very possibly impaired vascular systems of adults. They will, therefore, feel the cold less than adults and need to be dressed less, not more, heavily.

Of greater concern with kids is keeping them dry when they go outside, which means making sure their hands don't get soaked while building a snowman, their knees stay dry when they fall off their sleds for the tenth time, and their backsides are toasty even though they have been lying in the snow making angels by flailing their arms and legs.

Dress your kids in layers, particularly if you do not plan to keep an eagle eye on them every moment they are out. Not only will the layering work to keep them warm; it will also ensure that if they remove a garment or two because they feel hot and sweaty, they will still be dressed to some degree!

Pay particular attention to ventilation. An active child will sweat, and dampness from that source is as chilling as from any other. You can overdress your child and unwittingly make it colder going than if you had stuck to light layers with lots of avenues for evaporation of body moisture.

Watch for gaps in clothing. Choose bib-type snow pants instead of waist-high ones, tuck sweaters into elastic waistbands whenever possible, and make sure sleeve and mitten arrangements are invulnerable to riding up—wind, and more likely snow, will find a way to the skin if given half a chance. And do have a good supply of warm, dry mittens on hand; keeping a relay going so your children never go out with water-soaked mittens is probably the greatest favor you can do for them.

Although a hood would seem to be a desirable part of your child's outerwear, think twice before you make that hood down-filled. Among other things that the loft of down does is insulate against *noise*. What this means in practical terms is that your child may not hear car horns or other warning sounds if his or her ears are

---

Legwarmers for Kids

Here are two recycling projects for making legwarmers for kids.

Legwarmers from Mom's Sweater
1. Measure the length of your child's legs, as well as the width of calf and thigh.
2. Cut the sleeves of an old sweater at the shoulder seam. Cut away any excess fabric so sleeve matches in width the child's calf and thigh measurements, plus 1" (2.5 cm). Even off top edge, cutting away sleeve cap. Pin the edges together with right sides facing, and sew up seam with a tight zigzag machine stitch.
3. Cut two pieces of 1/2" (1.3 cm) wide, soft elastic as long as your child's thigh is wide, plus 1" (2.5 cm). Turn over

3/4" (1.9 cm) of the top edge of the sleeve to make a casing for the elastic, and stitch it closed by machine, leaving about 1" (2.5 cm) open for inserting the elastic. A safety pin fastened to one end of the elastic will make it easier to work it through the casing. Securely tack the ends of the elastic together and stitch closed the hole you left.
4. The cuff of the sweater will become the ankle part of the legwarmers, but energetic children particularly sometimes need a stirrup to keep the legwarmers securely inside their boots. Sew a piece of elastic to one edge, and a snap to the opposite edge, matching that one with the other half of the snap on the elastic.

Legwarmers from Worn-out Doctor Denton's
1. Cut off the feet and cut across the top of the pajama thigh right at the crotch seam. You will end up with two tubes slightly wider at the top than at the bottom.
2. Make a 3/4" (1.9 cm) casing for elastic at the top and bottom of each tube, leaving a small open space for threading the elastic. Measure the ankle and thigh tops of your child and add 1" (2.5 cm) to each measurement. Cut four lengths of soft 1/2" (1.3 cm) wide elastic to those measurements. Work the elastic through the casing with a safety pin, then tack the two ends of the elastic together and close up the casing hole. Repeat at top and bottom for each tube.

—J.M.

protected from the cold with a down-filled hood. A woolen hat covered with an unlined, waterproof hood is probably a better idea.

Rubber boots are the most practical footwear for children in the wintertime, but follow the principles outlined in the boot discussion above: Be sure there's lots of toe space; use at least two pairs of socks, one soft, light and absorbent, the other a good insulator; make sure the boots themselves are lined and that there are no gaping entrance points for snow, slush, and rain.

The things that seem to make children most miserable in cold weather are chapped skin and burning cold fingers and toes. Protect their skin from chapping—lips especially—with petroleum-based ointments; apply before they go outside and again at bedtime. If your child complains on returning home of painful burning in the extremities, do not panic. For one thing, the fact that it hurts is the best evidence that it's not frostbite, since frostbite is characterized by a loss of sensation. The burning feeling is caused by an infusion of warming blood in the chilled area and it's a sign that nature is doing its job. You can relieve the discomfort, however, with tepid water or a warm blanket. But emergency measures are not needed, and a cup of hot chocolate and an assurance that the feeling will soon go away are probably just as effective.

Keeping infants warm is another story. Babies are less active, particularly outdoors, than toddlers and older children, and they are not as acclimatized as those of us who have been through a few more winters. Nonetheless, that does not mean you should smother an infant in blankets and buntings, and you should certainly not keep him or her inside for the winter.

Infants should go outside in cold weather; it will help their bodies learn to adjust to heat and cold. The best time for a winter outing is midday. Remember that temperature is not the only factor; moist air is more chilling than dry air, and wind speed influences how cold the air feels. Dress the baby in a long-sleeved wool or cotton

Thin Ice

1. Don't go on questionable ice alone:
   1" (2.5 cm) of ice—keep off
   2" (5 cm) of ice—use caution
   3" (7.5 cm) of ice—okay for small groups
   4" (10 cm) of ice—okay
   6" (15 cm) of ice—okay for a snowmobile
   7" (17.5 cm) of ice—okay for most automobiles
   8" (20 cm) of ice—okay for a light truck or van
2. Slushy, dirty ice is not as safe as clear blue ice, and snowcovered ice may be weak because snow provides insulation. Rivers and streams have currents that may weaken the ice at any point, though the edge is likely to be more solid than the center.
3. If you fall on cracked ice, remain prone and roll away from the danger area.
4. Carry a long, light pole when you are traveling across ice. If the ice should start to crack, lie down spreadeagle, holding the pole between your outstretched arms and try to crawl away from the danger area. The pole will help to distribute your weight.
5. If you do fall through the ice, prop the pole across the opening and hang on so that you will not float away from the point of entry.
6. Do not press on the edge of already thin ice; extend your arms forward, kick your feet to an extended and nearly level swimming position, and work yourself forward on the ice. A knife or long, sharp nail, if you had the foresight to carry one, can be driven into the ice at arm's length to be used for anchorage. In extremely cold weather, if you put your wet sleeves and gloves on the ice they may freeze there, giving you emergency leverage. When you can, lift your leg out of the hole and roll toward safety. Shaded ice will be thicker than that exposed to sunlight, so if possible roll toward the shade.
7. To rescue someone else from a fall through the ice, remain prone if you must go on the ice yourself. Extend a rope, pole, ladder, or plank toward the victim. If possible, rescuers should wear skates or ice creepers to anchor themselves.
8. If there is no rescue material available, a human chain can be formed. Those forming the chain should lie on the ice, keeping their bodies rigid, so that the chain resembles a ladder. This chain should then be extended toward the victim; make sure that those on the edge of the water have adequate anchorage to pull the rest of the rescuers and victim out of danger.
9. The victim should get out of wet clothes as soon as possible and be taken to a warm area. If you are in the wilderness and there is little hope of getting into dry clothes quickly, rolling in soft, fluffy snow immediately may help to absorb some of the water before it penetrates clothing.

undershirt that attaches to the diaper and keeps the belly warm. Add a stretch suit, a zippered sweater, and as the outer layer a pram bag or snowsuit. If you are using a carriage, tuck a knitted, woven, or crocheted blanket around the baby. Or use a frontpack or Snugli. These slinglike contraptions are designed to accommodate a child from newborn to about three years old, and permit the sharing of body warmth since the baby is held close to the adult.

Dr. Spock admits that one of the hardest questions for a pediatrician to answer is how warm an infant should be kept. Babies and young children have an extra layer of fat, and at the same temperature children who are reasonably plump may need less covering than an adult. Most babies are overdressed, which inhibits their bodies' ability to adapt to changes in temperature. Also, babies' hands are often cool, so don't judge temperature by that. If a baby is genuinely cold, the face will lose color and he or she will start to fuss.

Indoors the most practical covering for a baby's crib is a synthetic blanket, since it is both warm and easily washed. A knitted shawl is a good covering for a very small baby who can be wrapped up securely in it. When the baby is bigger and kicks and moves around a lot, make sure that the blankets you get can be tucked into the mattress securely.

But as any mother and father knows, one of the joys of bedtime for young children lies in kicking off their blankets so they can get tucked in again. Little babies can be kept quite comfortable at night in stretch suits with "feet," or long cotton knitted nightgowns—if they are long enough, it will be hard for the baby to kick off the blankets. Some of these nightgows come with mittens on the ends of the sleeves, primarily to prevent babies from scratching themselves, though they might keep their hands a little warmer. (Most babies we know hated the mittens.)

When babies begin to crawl around in their cribs, and certainly after that wonderful day when they discover they can actually leave the crib on their own power (sometimes head first, but that's another story), blanket sleepers are a warm idea. Really one-piece pajamas that zip up the front, they are warm enough to take the place of both pajamas and blankets. They often have nonskid material on the bottoms of the feet, but that's not warm enough to insulate against really cold floors. When buying a blanket sleeper, check its thickness by pinching it; this, as always, is a good indication of warmth. Some sleepers have pile, as blankets do, and those with longer pile are likely to stand up better to repeated washings.

The Diving Reflex

Don't give up on a drowning victim too soon. Although ordinarily there is a four-minute limit before irreversible brain damage or death occurs, the survival time in water below 21° C (70° F) can stretch this limit to longer than thirty minutes.

Even if there is no detectable pulse or heart beat, the pupils of the eyes are fixed and dilated, and the skin blue, resuscitation may be possible. Artificial respiration or cardiopulmonary resuscitation (practiced by a trained person) should be started immediately and continued until there is absolutely no hope of revival. One victim received CPR for two hours and breathing assistance in a hospital for thirteen hours before regaining consciousness. There was no brain damage.

Victims of cold water submersion have been saved by their own body's "diving reflex." Mammals such as porpoises and seals take advantage of this reflex every day of their lives. Triggered by the holding of breath and the cold water on the body, the reflex operates for humans also. The cold water reduces the body's need for oxygen; the reflex slows the heartbeat and therefore the flow of blood to skin, muscles, and other tissues relatively resistant to oxygen deprivation. The remaining oxygenated blood is sent to the heart and brain, which would die if their supply were cut off.

Scientists think that the human body may experience this reflex to some extent when a baby passes through the birth canal and is temporarily deprived of oxygen. In fact, it has been recognized that the reflex is stronger and more swiftly triggered the younger the submersion victim is.

When bathing an infant in cold weather, do it in a warm, draft-free room. The kitchen might be the best place in that respect. Water temperature should be about 35° C (95° F), and obviously you should dry the baby thoroughly after the bath.

The rest of it is really just common sense. Colds and flu seem to flourish in the winter, and children, particularly of school age, are ready targets. Keeping them warm turns out to be a less vital prevention than keeping them well rested and well fed. But keeping them warm will contribute greatly to keeping them happy.

In general, dressing to keep warm at home or outside is a matter of having the proper clothes *and* knowing how to wear them. But if it's a cold, cruel world outside, there's no reason for your home to be that way. Let's look now at some ways to make your surroundings cozy and inviting no matter what it's doing outside.

Decorating for Comfort

You can turn any room into a warm retreat against the howling winds outside. Your approach to decorating can be practical, fanciful, or a little of both. The task can be ambitious—carpet a wall, reupholster a couch—or simple—sew up some throw pillows or an instant homemade quilt. Whatever you choose to do, there are some basic principles to keep in mind.

Some types of furniture are cozier than others. A big, soft sofa with lots of pillows will look and feel warmer than a wood-frame-and-canvas couch. A wing chair invites one to curl up in it, and the extended sides add protection against drafts. But you can decorate to your taste and be warm, too, and the investment in winter clothing for your house does not have to be enormous. The simplest trick is to add pillows, afghans, blankets, and throws. You can make or buy quilted or padded slipcovers for winter. (Store them away in the summer when a cooler look and feel is needed.)

When you choose upholstery or slipcover fabrics remember that plushes (velvet, corduroy, suedecloth) will trap the warm air generated by your body and give you a warm feeling. Plastics and vinyls are hard to warm up to, but a rich, soft leather might be a welcome change from woven fabrics.

And do consider rearranging the room. Furniture drawn close together will make sitting in the room warmer and more comfortable. A fireplace, whether built-in or freestanding, is a good focal point for the room. Keep chairs and sofas away from windows, doors, and outside walls, where drafts and heat loss are more likely to occur.

Make sure that radiators and other heat sources are not blocked by bulky pieces of furniture. Even radiator covers cut down on heat efficiency. If you do use them, make sure they allow adequate ventilation; check particularly that the grille holes, which should comprise 75 percent of the surface, have not become clogged with paint. Drill more holes and/or knock out the paint if necessary to achieve a 75 percent open-work surface. You can increase the amount of heat reflected into the room by placing a sheet of aluminum foil behind the radiator, the shinier side facing the room.

Color can be used to make a room seem warmer. The red-orange-yellow end of the spectrum, the "warm" colors, can serve on walls, floors, even ceilings, as accent points or vast stretches of color. The cool blue-green-purple range should be avoided or counteracted wherever possible.

If a high ceiling makes the room feel cold, try painting it a shade darker than the rest of the room; that will tend to make the ceiling look closer and, in that way, make the room cozier. If you have molding a foot or so from the ceiling, paint the top section of the wall the same color as the ceiling, rather than the color of the wall below the molding; this too will bring the ceiling "down."

Covering your walls will augment their insulation and can add to the general feeling of warmth. Cork tile gives a toasty look, insulates, and is an effective acoustical barrier as well. Velvet wall covering will make you feel as though you are surrounded by a warm blanket; flannel wall covering, a little less elegant perhaps, will serve the same purpose. Another choice is to carpet the wall. The installation should be done by a professional, because carpeting is very heavy and difficult to manipulate. Or you can buy self-stick carpet tile, which is easy to put up by yourself.

Do, if possible, carpet the floor. Carpets are warmer to the touch and offer more insulation than tile, wood, slate, or resilient floor coverings. Wall-to-wall carpets, laid over a cushion or pad, have greater insulating qualities than room-size or area rugs. Generally, the thicker and denser the carpet, the greater the insulation value. Wool and acrylic fibers provide the best insulation; nylon is the least effective.

Start with a carpet pad, either fiber and jute or foam rubber.

Do-it-yourself Scatter Rugs

Your home is probably filled with materials you can use to make a warm, durable rug that will add a note of warmth to the furnishings; and rug-making in front of a fire might be the perfect activity for a long, cold winter night. The rugs described below require no special skills and no special materials.

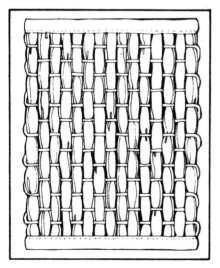

Simple Box or Frame Rugs:
Make a loom using either an old fruit crate or a frame made of white pine 2'-x-2's (6m-x-6m) (reinforced at the corners with metal angles). The size of the loom will determine the size of the finished rug.
1. Mark off 1/2" (1.3 cm) intervals along two opposite sides of the crate or frame.
2. Drive in ten penny nails at each mark.

3. Using a single length of heavy cotton string (either white or colored), thread the loom from top to bottom by drawing the string around the nails. This will form the warp of the rug.
4. For the weft, or filling material, you can use old clothing cut into strips, nylon stockings, old blankets or curtains, and various other fabric scraps. It is best to use the same type of fabric (all nylon, or all woolen) so that the rug will wear evenly and cleaning requirements will be consistent.
5. Cut the weft strips about 6" (15 cm) longer than the width of the loom, to leave a 3" (7.5 cm) fringe on each side of the rug. Or, if you do not want a fringe, use a continuous strip of weft fabric, sewing pieces together before you start to weave if necessary.
6. Bring the weft material over and under the stretched warp cord—over the first thread, under the second, over the third, and so on. The second weft row should be the opposite of the first—under the first thread, over the second, under the third, and so on. All odd rows will match the first, all even rows will match the second.
7. When you have filled in the entire piece, tie the warp threads with a length of cord to prevent the woven strips from slipping. Slip knots or half-hitches along the ends of the rug should be sufficient.
8. Remove the rug from the loom.

Braided Rugs:
1. Cut or tear strips of fabric into 2½"-3½"- (6.3 - 9 cm-) wide pieces, making them as long as possible. It is best to use one type of compatible fabric for a rug.
2. Fold the strips so that the rough edges meet in the center of the wrong side of the fabric, and then bring the folded edges together. Since it is important to keep the rough edges inside, you might

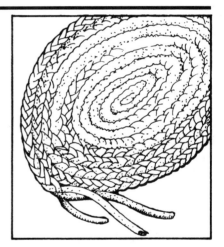

want to iron each strip; to save time, however, you can usually preserve the folds by rolling each strip into a tight coil as you prepare it.
3. To start the braid, pin three strips together with a safety pin.
4. Begin braiding by folding the right strand over the center strand, and the left strand over that. Keep the folded edges toward the center as you braid.
5. When a strand ends, simply attach another by machine-stitching a bias seam.
6. Assemble the rug as you braid. Coil the braided strands, and lace the loop of one braid to the loop of the braid that falls next to it in the coil. A blunt needle, threaded with double strand of carpet thread or string will provide the strongest connection.
7. As you approach the last 15"- 20" (27.5 cm - 50 cm) of the rug, cut the braiding strips thinner, so that the braid will eventually become thin enough to be attached with an overhand stitch to the body of the rug.

Padding will not only increase the life of your carpet, but will add an extra layer of insulation. Wool carpeting provides the warmest feel underfoot, though less expensive acrylics can come close. The closer the weave of a carpet, the warmer it will be. Plush, fur, shag, and rya-style carpets make any room look inviting. In an old house, extend the carpeting up over the molding to block any drafts that might seep through the cracks between the floor and the molding.

Other floor covering, though it does not offer the direct warmth of carpeting, will help insulate your floors. Various semihard coverings are possible. Cork tile has good insulation value, is warm, resilient, and quiet. Vinyl flooring, either in sheets or tiles, is as expensive as medium-grade carpeting, but makes more sense in kitchens and other work areas. Good quality vinyl covering usually has cushioned backing, making it warm and comfortable to walk on. (Unbacked vinyl will be cold.) Less expensive than vinyl, a thick linoleum is warm underfoot.

Unadorned wood floors do not generally get too cold, and they do have some insulation value. Besides, a polished hardwood floor can be beautiful indeed, and if you are loath to hide one beneath wall-to-wall carpeting, use area rugs scattered about where they will be most appreciated—at the foot of a favorite chair or in front of the fireplace. As stunning as they can be, stone, ceramic tile, and brick floors are a chilling proposition. Area rugs are a great help in this situation. If you do have floors of stone or brick, a coat of polyurethane sealer will cut down on dampness.

Drafts from under doors can be blocked with a sand-filled fabric tube laid on the threshhold.

Window Treatments

Windows have considerable effect on heat loss in winter and heat gain in summer. In a recent study, it was found that the use of a light-color opaque roller shade mounted inside the frame reduced heat loss through the window by 24 to 31 percent during the winter. In summer it could provide a savings of 21 cents of each air-conditioning dollar. The second best insulator, reducing heat flow approximately 21 percent, was lined-glass-fiber draperies, sealed on the sides, top, and bottom, and pinned together at the center closure.

The method of draping a window has considerably more effect on heat flow than the fabric used. If your home has heat registers,

---

Draft Stopper

A frequent sight in New England houses, the threshold draft stopper is an ingenious device. You can find them in many mail order catalogues, but it is quite easy to make one yourself.

Cut a piece of fabric about 8" (20 cm) wide and 3" (7.5 cm) longer than the door in question. Fold it in half lengthwise, right sides together, and make a 1/2" (1.3 cm) seam. Stitch one end, leaving the other open. Turn the tube through the open end, fill it with sand, and stitch the opening closed.

Lay the draft stopper at the threshold on the inside of any door; under-the-door drafts will be blocked. When not in use, the draft stopper can be hung over the doorknob.

draperies should hang at least 8" (20 cm) above the heat source, or metal deflectors can be used to direct heat into a room. For best efficiency with radiator or baseboard heat, draperies should hang a minimum of 4" (10 cm) above the heat source.

Since draperies are usually extended 3" (7.5 cm) from a wall, closed draperies can form a draft tunnel, with warm air being drawn over the top of the draperies and cooled on the glass, then flowing back into the room under the bottom of the draperies. One way to correct this situation is to construct a closed-top cornice or valance of wood or other material to fit over the top of the draperies. The drapery should be in contact with the valance and should be secured at the bottom.

Any drapery will hold warmth more effectively if it is nailed or stapled down to the sides of the window frame. The sides can also be nailed to separate furring strips attached to the frame. The furring strips fit snugly behind a couple of wooden blocks nailed to the sill.

Another option is shutters, either inside or outside the house. Single-paneled outside shutters can be closed when there is a fierce wind, and they will help keep the warmth built up during the day inside the house at night. Awning shutters are another outdoor possibility; they have a hinge halfway up so the shutter can be opened partially for ventilation and light. The most common design for inside shutters has hinges in two or three sections so the shutters can be folded out of the way when opened. These are usually louvered to let in some light.

Whatever you use, keep window coverings closed at night and on overcast days; open them when it is sunny to get the benefits of solar heat.

Other Rooms

People will congregate in warm rooms. Once upon a time, the kitchen was the center of the household for this very reason. With its blazing wood fire and the tantalizing smell of cooking food, the kitchen was a natural haven. Although most of us do not cook on wood stoves any more, modern-day kitchen appliances do give off heat. Not the least heat generating is the refrigerator. It is the one appliance that runs all the time, and although its inside is frosty cold, the exhaust vent is a source of very warm air. Take advantage of that heat to warm the kitchen. The aroma of good food will always evoke a warm response. Try to re-create the welcoming atmosphere of an old-fashioned kitchen where you live. The key is simplicity—natural materials (wood, brick, stone) and natural colors. A display of cooking utensils, baskets of fruits and vegetables, an herb garden or other plants will all contribute to the homey feeling.

After a long cold day outside, a warm bath provides an easy way to relax and get warmed up before joining the group in the kitchen. A leisurely bath, scented with oils or bath crystals, if you like, will restore your circulation and your good spirits as well. The bathroom can be the warmest room in the house; even if yours is tiled from floor to ceiling and filled with frosty white porcelain fixtures, there are some simple ways to make the room warmer. If you're ready for complete redecoration, consider covering floors, walls, and even the tub (outside, of course) with polyurethane-sealed cork tiles. Wood is also warmer than tile and can be treated with epoxy resin to make it waterproof. Vinyl wallpaper, well suited to the damp milieu of the bath, comes in a dazzling array of patterns and colors. Orange or yellow formica-covered counters mean the look of continual sunshine in the bathroom. Thick, velvety towels, matching or contrasting plush carpeting (make sure it is backed with a nonrot synthetic or rubber, not jute), pictures on the walls, and pretty jars to hold your bathroom necessities can all add to a feeling of warmth. Plants thrive in the steamy atmosphere of most bathrooms and will add a touch of the tropics even in a starkly tiled room. A heated towel rack can supplement the heating system in a small, underheated bathroom; and think of the luxury of cuddling in a warm towel after a shower or bath! If the room is adequeately heated, lay your towel and robe over the radiator while you bathe.

Bedrooms

In the dead of winter, who is immune to the occasional overwhelming urge to spend the day in bed? In the morning, your bed is warm and toasty because you have spent the whole night warming it—did you know the average adult gives off between 350–400 BTUs per hour? If your thermostat has been set low for the night, the rest of your bedroom may seem icy in comparison to your bed.

One way to make the bedroom warmer, and incidentally more intimate, is to hang a decorative fabric tent from the ceiling. This will effectively reduce the amount of space that needs to be warmed, and by insulating the ceiling, will slow down the loss of heat upward. The old canopy bed idea is another way to bring tenting inside. If side curtains are hung from the ceiling or canopy frame to enclose the bed entirely in fabric, real warmth can be enjoyed even in a very cold room, largely because human body heat is thereby confined by the bed draperies. Similarly, heavy draperies on the windows will help retain heat in the room.

Cold floors are never inviting in the winter, but stepping on one immediately upon waking in the morning is somehow the least appealing notion of all. If you choose not to carpet your entire bedroom, at least use scatter rugs near the bed, or make a carpeted platform for your mattress, leaving enough room around the mattress to put your feet on when you roll out of bed in the morning.

Early settlers in America were well aware of the need for warm bed coverings. In the 1630s, both Massachusetts Bay Colony governor John Winthrop and Maryland governor Lord Baltimore wrote back home warning those planning to make the voyage to the cold New World to bring "rugges" for their bedding. The "rugges" they referred to were bed coverings made of coarse woolen homespun thickly decorated with brightly colored embroidery. Rugges were replaced in the eighteenth century by quilts, which came to England via the Orient, where British soldiers discovered that Turks wore several layers of quilted fabric under their armor. The Turks were interested in quilting as effective padding against armor chafe, but the British soon figured out that as bed and body covering, quilts were blessedly warm.

The earliest quilts were made of linsey-woolsey, consisting of a bottom layer of linen and a top layer of wool, with a pad of raw, unprocessed wool in between. The three layers were held together with small tight stitches. By the late eighteenth century, the patchwork quilt had made its appearance. More colorful than the plain linsey-woolsey, it too relied for warmth on three layers sewn together. None of these bed coverings could be used alone in the colder northern climates, and inventories of houses in the eighteenth century all indicate an abundance of blankets, quilts and coverlets.

Today, our houses are more effectively heated and insulated, so we do not need the many layers used by the early colonists. Nonetheless, a wise selection of bed clothing will add to bedtime comfort, particularly if you turn down your heat at night.

Once again, layering is the key. Beginning at mattress level, a quilted mattress cover serves double duty. In addition to protecting the mattress, it provides a layer of insulation. If it's especially chilly, add to that a light- to medium-weight blanket before you put on the bottom sheet.

Even toastier, though operating costs should be considered, are electric bed warmers, mattress covers wired like electric blankets to place the warmth under rather than over you. According to Patented Products, currently the only source of this item, heat rises through the sheet and collects under the blanket, whereas electric blankets radiate heat outward as well as in toward the body. Bed warmers, they claim, use half the electricity to provide the same amount of warmth as electric blankets do.

After mattress covers come sheets. Today's easy-care polyester sheets come in a splendid array of colors and patterns to cheer and add touches of warming color to the bedroom. But no matter whether they are snowy white or richly colored, cold, crisp sheets are less than inviting on a wintry night. One solution is the old-fashioned blanket sheet, made of thickly woven 100 percent cotton flannel, an item that used to be found in every country house in New England. They are much warmer than conventional sheeting—really somewhere between a light blanket and a sheet—and are available in a wide variety of colors. Several bedclothing manufacturers have come out with "thermal" sheets made of brushed nylon knit or cotton and polyester flannel. Machine washable and nonstatic, they have the cuddly feel of flannel pajamas. Most department stores carry these.

The top, or outer, layer of your bed can consist of blanket, quilt, featherbed, comforter, duvet or puff. The choice today is wider than it has ever been; and shopping for a bed covering can be very confusing. Because this layer can represent a considerable investment, let's try to clarify what some of the choices are really about.

When it comes to buying blankets, you are faced with a choice of three general types: conventional woven, thermal, and nylon-flocked polyurethane-core blankets. The warming powers of a blanket are dependent on its bulk and nap, not on its weight. Once again, the ability to trap still, warm air is what counts. When you shop for a blanket, pay attention to how quickly it returns to maximum thickness after it has been compressed. That is the easiest way to judge loft in a blanket. A conventional-weave blanket made of synthetic fibers may be warmer than an all-wool or wool-blend blanket of comparable weight. On the other hand, wool retains its nap longer than synthetics do, so a wool blanket might be a wise investment because, in effect, it stays warmer longer.

Thermal blankets, like thermal underwear, have a honeycomb or waffle weave. Lighter in weight than conventional weaves, they are advertised as all-weather blankets. The open weave, which allows body heat to escape during warm weather, is a disadvantage if you are concerned primarily with keeping warm. Although any thermal blanket used alone is not as warm as a conventional blanket (just as a thermal undershirt worn alone is not as warm as a sweater), its insulating properties can be increased by enveloping it in a blanket cover (or even sandwiching it between two sheets). Napped thermal blankets are slightly warmer than unnapped ones.

Nylon-flocked blankets have a polyurethane foam core to both sides of which nylon fibers are electrostatically bonded. The core serves to retain body heat. These blankets are very warm and soft to the touch. They tend to wear out quickly, however, and once the material has worn thin, most of its insulation ability is gone. Because the principle of layering is as valid in bedclothing as it is in clothing for people, a couple of lighter-weight blankets will insulate more effectively than one heavy blanket. But if you find the weight/warmth ratio of either too heavy a load to bear, a reasonable solution is the electric blanket. If properly cared for, electric blankets are perfectly safe and cozily warm. True, it costs between five and twenty-five cents a night to run one (depending on the blanket and your local power rates), but many people find it easier to endure a lower thermostat setting if they can sleep in an electrically heated bed.

Electric-blanket technology has advanced a great distance since the early days when they were considered unsafe and possibly unhealthy. Today's electric blankets have heating coils sandwiched between two blanket layers. The wires are well-insulated and hard to damage. One thing that will damage the insulation is dry cleaning, so always hand or machine wash an electric blanket according to manufacturer's instructions.

For cozy sleeping without electricity or the weight of blanket layers, consider filled comforters. Feather or down comforters are an old-fashioned solution that is still practical today. Less expensive than down- or down-and-feather-filled comforters are those filled with polyester batting. The range of styles and prices is very wide. You can find comforters in solid colors or a variety of prints, often to match curtains and sheets; they come covered in cotton, cotton-polyester blends, or shiny satin and taffeta finishes. Do make sure your comforter will stay put; as luxurious as the satiny ones are, they are a good deal more slippery than cotton-covered ones. Another possibility is to buy a comforter covered in plain nylon taffeta and make a cover for it. (See the directions for the blanket cover below).

Down comforters are certainly the champagne of bed covers, but if your budget is beer, don't despair. Several of the outerwear-kit manufacturers also make comforter kits in crib, twin, double, and king sizes. Each kit includes down filling, nylon taffeta yardage, and instructions on how to make the cover.

A variation of the comforter, which generally has an overhang like a bedspread, is the European duvet or puff, which is just about the size of the top of the bed. Duvets and puffs are generally not

---

Blanket or Comforter Cover

You can buy a blanket cover to complement or to match your sheets, or you can make one in no time. All you need are two sheets and seven snaps.

On the wrong side of one sheet, sew a snap near, but not at, each corner. Sew the other half of each snap to the corresponding point on the blanket. If your blanket is a standard size, it will be as wide as the sheet but not as long. Take 1/2" (1.3 cm) seam allowance into account for the length before sewing on snaps. For the width, allow for overlap at the edges or, if you prefer, trim the long side of the blanket by 1½" (4 cm).

Sew the two sheets together, right sides facing, along the two sides and at one end. Turn the case rightside out. Sew three

snaps to either side of the open edge so the cover can be closed when the blanket is inside. (You can also use Velcro to close edge, or three ties made from ribbon or strips of sheet).

Slip the blanket into the cover and fasten the corner snaps to hold the blanket in place.

If you intend to use the cover for a comforter instead of a blanket, you will probably need sheets a size larger than the comforter to accommodate its greater loft. Measure first to be sure.

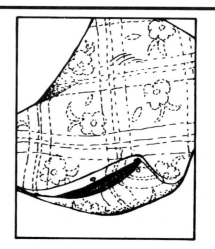

quilted in channels like most other comforters are, though the names are frequently interchangeable; what one manufacturer calls a duvet may be your idea of a quilt.

Polyester-filled comforters are washable; down-filled ones may or may not be—check the label before you buy. Dry cleaning is, as with down outerwear, a tricky business. When in doubt, do not do it. The most practical solution is to keep comforters in removeable washable covers. The comforter itself need only be aired out from time to time; the cover can be washed with ease. If you have a clothes-line, so much the better; otherwise, just shake out the comforter in clean, fresh air.

Let's suppose that you are now well-equipped for bedtime comfort, but the thought of slipping between crisp, cold sheets does not appeal to you. If you have an electric blanket, turn it on a short time before you go to bed. There are other ways to warm the sheets, especially at the foot of the bed where the cold is most unpleasant to encounter. Try a heating pad or a hot-water bottle. If you are lucky enough to find an antique warming pan, you can fill it with hot coals from your dying living room fire and warm the lower part of your bed that way; or wrap in flannel a brick previously warmed in the fireplace and place it under the covers at the foot of your bed. Only the hot-water bottle is safe to spend the night with. If an electric heating pad gets squashed under any part of your body for a long time, a severe burn can result. Heating pads, by the way, should never be used by people with impaired circulation or diabetes; nor are they recommended for the elderly or for anyone taking cortisone.

How you dress for bed makes a big difference in how warmly you sleep. Pajamas are the unisex item of choice; nightgowns and nightshirts simply will not keep you reliably covered and warm. Choose loose-fitting styles in brushed cotton, flannel, or knits. Thermal underwear and sweatsuits are two other warm sleepwear possibilities. Socks are a good idea if your feet tend to feel cold at night, and a

---

Two Quick Quilts

Quilt 1

If you have the patience, consider making a patchwork top for your quilt; the bottom (or bed side) can be made of a sheet. Or make both sides from sheets. The quilt should be 20"- 27" (50 cm - 67.5 cm) wider than the bed. If you use sheets, the normal overhang should be adequate. The only other thing you need is polyster quilting batts, available in most needlework shops or notions departments.

Center the batts on the wrong side of one sheet and baste them in place; then pin the right side of the other sheet to the right side of the first. (The batts will be on top.) Stitch around the assemblage, leaving 20" (50 cm) open on one side to permit turning the quilt.

Turn the quilt rightside out and draw a quilting pattern on the cover lightly with pencil or tailor's chalk. Machine or hand stitch through all three layers.

Quilt 2

Working with polyester batting and fabric large enough for even a single bed can be tricky simply because of the bulkiness of the materials you are working with. If you would like to try a quicker and simpler quilt, here's one made with an old (or new) mattress pad—the flat, uncontoured sort.

Because that style is only as wide as the mattress top, you will need a double-bed mattress pad for a twin bed, a queen-size mattress pad for a double bed, etc. If you have a king-size bed, you will have to sew two smaller mattress pads together. If you use a new pad, wash it before making the quilt to eliminate shrinkage of the finished piece.

Sandwich the pad between two sheets that are the same size as the mattress pad. Sew around all four edges to anchor the three layers, then add decorative stitching across the top layer through the padding to the bottom. Bind all four edges with bias tape in a contrasting color.

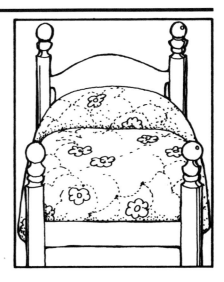

night cap (don't laugh) will prevent heat loss through radiation from the top of your head.

Keep a bathrobe at the end of your bed or otherwise within arm's reach, and be sure a pair of slippers is close by. It will make it just a bit easier to contemplate getting out of that nice warm bed come morning.

Sleeping Bags

Even if you are not a camping family, having a couple of sleeping bags in your home makes good sense. They are an easy way to accommodate overnight guests, and most kids will consider them more fun to sleep in than a blanket or quilt. More important, they can be life savers if your home is without heat during a weather emergency.

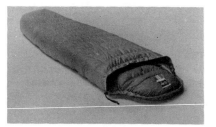

Down-filled sleeping bags are generally warmest by weight; bags filled with synthetic fiber are less expensive and nonallergenic. Most sleeping-bag manufacturers indicate the temperature range within which a particular bag will provide adequate warmth. There is, however, no standard formula for determining this so-called comfort range, so take the label claim with a grain of salt.

In addition to the quantity and the quality of the filling, how the bag is constructed can affect how warm it is. This is of greater importance outdoors, but if your house is likely to be drafty as well as cold, you should watch for certain construction features. Bags with stitching through all three layers (liner, filling, and outer shell) have cold spots along every stitched row. Baffle construction, which employs strips of nylon or mesh to connect the inner and outer shells and channel the filling, makes for a more uniformly warm bag.

There are four baffle styles. The simplest and least effective is the "box tube," which consists of parallel strips running from the inner to outer shell. The "slant tube" is an improvement on the box tube, since the filling of one compartment will compensate for the thin corners of the adjoining compartment. Overlapping "v-tube" construction is satisfactory even in subzero temperatures. Like triangles paired head to foot, the tubes provide a uniform layer of insulation as the thin part of the one triangle is compensated for by the thick part of its next-door neighbor. Laminated or double-quilted construction resembels a brick wall with two layers of baffles, each division between the baffles on one layer protected by the total area of the other layer's baffle.

Sleeping bags come in three basic shapes. The rectangular, or "sack," is the roomiest. It allows you full freedom to stretch and turn while you sleep. The advantage of a rectangular sleeping bag is that it can be unzipped all the way so it can be used as a comforter too. The mummy bag conforms to the contours of your body, and so is much warmer than the rectangular shape, which leaves a lot of air space between you and the bag. Old-fashioned mummy bags were quite difficult to get into because they had no zippers; now that they have zippers, no one should have the least bit of difficulty slipping into these down-filled cocoons. Although mummies do inhibit your movements, a little practice will soon have you turning with your bag without falling off the bed or air mattress. Most mummy bags have hoods attached, and some have a down-stuffed collar to prevent drafts at the shoulders. Those that don't have hoods usually do have a zipper for attaching a separately purchased hood. A hybrid form, the barrel bag, is somewhere between the rectangle and the mummy. Fitted at the top and bottom like the mummy, it expands at the waist-to-knee area. This feature might be appreciated by the mummy lover who sleeps in the fetal position.

Consider sleeping bags instead of quilts or blankets for your child's winter bedcovering. You will not have to worry about covers getting kicked off in the middle of the night, and most children will find them fun. They are practical only after a child is really toilet trained, of course; washing a sleeping bag in the middle of the night is no treat. You can generally find sleeping bags in sizes scaled down for children, though they are almost exclusively made with synthetic fillers. For cold emergencies, one of the best bags for a child is the add-on kit from Frostline. This down-filled bag "grows" with your child because you can purchase separate 12" (30 cm) sections to be added on as needed; the basic bag is 42" (105 cm) long. Other suppliers now carry PolarGuard bags for people under 5' (1.5 m). If your child is very small and you would prefer not to have him disappear into the depths of even a child-sized sleeping bag, consider a mountain climber's footsack. Used by climbers in conjunction with parkas when they have to sleep in weird positions and places, it is the perfect size for a child from six months to five years, and it is very warm. If you prefer down or if you have a closetful of adult-sized bags and don't intend to buy any smaller ones, you can tie off one at the middle to prevent your child from slipping down. In any event, sleeping bags are breatheable, so you don't have to be terribly concerned with a full-length bag, no matter what it's filled with.

Most people prefer bags with zippers. Nylon zippers are best, since metal zippers tend to break, freeze, or jam. The zipper should be covered on the inside of the bag by an insulating flap running the whole length of the zipper to prevent drafts. The flap should be suspended from the inside shell, not merely sewn onto the zipper edge, where there would be a cold line down the stitching. Two-way zippers, opening at the top for easy access and at the bottom in case your feet get too warm and you want extra ventilation, are a nice feature. Those of you who like to sleep in twos can buy bags that zip together—but be sure you've got one with a right-hand and the other with a left-hand zipper so that they do in fact go together!

Man's Best Friend, et al.

The other members of your household, dogs and cats, are generally well-prepared for winter. With their thicker winter coats and their natural instincts, they are less dependent on outside help than

---

Storm Warning: Supplies and Checks *Before the Storm Comes*

1. Check to make sure that you have adequate fuel. A supply truck may not be able to reach you for many days.
2. Stock up on wood for your fireplace in case power should fail.
3. Have a battery-operated radio or television. A two-way radio or CB is even better. This may be your only contact with the outside world for several days.
4. Make sure your first-aid kit is up to date, and that there is an adequate supply of any medicine you need regularly.
5. Have flashlights, candles, kerosene or gas lamps on hand in a readily accessible place. Make sure that there is enough kerosene or gas to fuel the lamps. 1½ pts (.7 l) of kerosene will light one lamp for about eight hours; 1 qt. (1 l) of gasoline will shine for nine hours.
6. Have your larder well-stocked with packaged goods: milk (canned, evaporated, nonfat dry milk), canned meat, poultry, fish, pasta products, canned fruits and fruit juices, cereal, fats that don't require refrigeration, coffee, tea, bouillon cubes. Make sure that some of this is ready-to-eat because a power failure may mean no way to cook.
7. Emergency cooking facilities—a camp stove, cans of sterno, a fondue pot— are a good idea in case of power failure.
8. Store some water in plastic containers for drinking, cooking, washing, in case pipes freeze.
9. Keep snow-clearance equipment (shovels or snow-blowers) in an accessible place, and spray them with silicone so that snow won't stick.
10. Keep rock salt or other snow-melting chemicals on hand.
11. Store enough extra blankets or sleeping bags to keep the whole household warm in case of a power failure.

---

mere human beings are. All the same, there are a handful of things to watch for during cold weather.

Dogs or cats, especially ones that spend a lot of time outside, must be well-fed in the winter. They need the extra calories to keep warm. The easiest way to achieve this is to add an extra meal to their normal regimen.

Dogs and cats are susceptible to colds and bronchial infections, and veterinarians believe that the time of greatest risk is when the animal is wet. For this reason you should consider eliminating baths from your dog's winter routine. (In any case, bathing robs your dog's skin of natural oils, necessary in winter; regular brushing should be enough to keep him clean and his coat healthy.) Be sure to dry the dog well after a bath, if you insist on bathing him. Cats will generally dry themselves off if they get wet from being out in the snow and rain or for any other reason but watch your cat carefully and assist with a brisk towelling if necessary.

Dogs and cats that spend most of the winter indoors may begin to shed, especially if they spend time in front of a fireplace or radiator. Brisk walks (for dogs) and a good brushing to stimulate the circulation and skin oils will help with this problem. Extra oil or fat in the diet will also keep the skin oilier, making the coat less likely to dry out.

Dogs, and their owners, may bridle at the idea of coats or sweaters, but if you live in the city and particularly if you have a short-haired dog, you would do well to give them a chance. Dogs walked on leashes do not have an opportunity to warm themselves with a brisk run; and if such a dog lives in a well- to overheated apartment with only twice daily forays into the cold, his natural coat may not be thick enough to protect him outside. Dog coats and sweaters should cover the chest and stomach as well as the back and shoulders.

If a dog has romped in the snow or walked on a snowy street, chunks of ice can form between the paws and in other places where the hair gets wet and then freezes. This can be quite painful. If you keep the hair on the paws clipped short, this is less likely to happen. A quick way of melting the ice is to use an electric hair dryer, but make sure that the heat is set on low. Salt is another problem, especially for city dogs. Washing your dog's feet with clear water when you get home is a must. If your dog seems to suffer from the salt— watch for cracked foot pads that look raw—consider buying rubber boots for him. You might run into some resistance from your pet, but give it a try if you like.

Newborn dogs and cats, no matter the season, are quite susceptible to chills. A nursing mother will take care of this to some extent; but you can help. Cover a hot-water bottle with a towel and place it with the puppies or kittens in their box. Remember to keep the water at a constant warmish temperature.

Cats will seek out the warmest spot in the house for their catnaps. Perhaps it will be a patch of sunlight on the floor of an east-facing room in the morning or a corner near the radiator later in the day. Enforcing a particular sleeping area will in all likelihood be futile if your cat doesn't think the place you have chosen is warm enough.

Your dog's sleeping quarters should be dry, warm, and free from drafts. An outdoor house should be sheltered from wind and rain. It is a good idea to raise it a few inches off the ground to prevent water seepage. The entrance should face south, toward maximum sunshine and away from the cold north wind. Cover the door with a canvas flap for wind and snow protection. No dog should sleep on a bare floor either in his house or yours. Give him an old blanket or rug. For inside buy or build a bed that is raised slightly off the floor. Make sure that the bedding remains clean and dry.

Perhaps the most beneficial arrangement for you and your pets is to allow them to sleep on your bed, sharing BTUs with you happily through the night. That's one way of taking care of your creature comforts!

Useful Addresses

If you do not have a large, well-supplied sporting goods or camping outlet near you, many manufacturers and suppliers of outerwear and other cold-weather gear have mail-order services. Indeed, many of them do business almost exclusively by mail. Here is a partial listing; catalogs are available from all.

Eddie Bauer
PO Box 3700
Seattle, WA 98124
(sleeping bags, outdoor clothing, tents, comforters)

L . L . Bean
Freeport, ME 04033
(outdoor clothing, boots, tents, sleeping bags, camping gear)

Eastern Mountain Sports
1041 Commonwealth Avenue
Boston, MA 02215
(outdoor clothing, tents, sleeping bags)

Frostline Outdoor Equipment
PO Box 1378
Boulder, CO 80302
(sleeping bags, outdoor clothing, tents, kits)

Garnet Hill
Box 262
Franconia, NH 03580
(blanket sheets)

Holubar Mountaineering
Box 7
Boulder, CO 80306
(comforters, sleeping bags, outdoor clothing, kits)

Patented Products
Danville, OH 43014
(electric bedwarmers)

Sierra Designs
247 Fourth Street
Oakland, CA 94607
(outdoor clothing, sleeping bags, tents)

Ski Hut
PO Box 309, 1615 University Avenue
Berkeley, CA 94701
(outdoor clothing, tents, sleeping bags, camping gear)

Stephenson
RFD 4, Box 398
Gifford, NH 03246
(tents, sleeping bags)

The Vermont Country Store
Weston, VT 05161
(outdoor clothing, blanket sheets)

Cold Weather
Cooking and Eating

Every now and again, there comes a good crisp day in the middle of winter when I feel as if I cannot stand to stay inside any longer. The sun sparkles brilliantly on a fresh fall of snow; sometimes the air is filled with tiny snowflakes that seem to crystallize out of nowhere. Indoors, the house is full of people hunting for socks and mittens, and then it is suddenly empty. I sit for a few moments, basking in the silence. I think about things I want to do: go for a walk, do a little sliding with the kids, dig out the root cellar. But it's cold out, bitingly, bitterly cold; we'll all be ravenous today. So I head for the stove.

We have already had a substantial breakfast. In addition to our usual eggs and scones, we've had a pan of home-fried potatoes and, as an afterthought, a jar of applesauce. That'll hold them until noon, I think, stepping outside to get a pint of beef stock from the "freezer"— a plastic bucket sunk in the snowdrift under the woodshed eaves. I've had some dried beans soaking overnight and I quickly assemble the beginnings of minestrone soup, setting it to simmer away the morning on the back of the wood stove. Then I bank the stove chockful of hardwood and hunt up my own mittens.

The root cellar is some kind of miracle. I shovel down two or three feet through the snow to a square wooden door set in the slope of the hillside. (I admit the system could be better, but it serves.) I pull back the canvas, lift the door, and presto! There is the brown earth, barrels of root vegetables, apples, cabbages. All around me, the world is flat and white and frozen; but here, under the snow, lies all our winter treasure, fresh and delicious as the day it was picked. I take a great fat carrot, rub it clean on my mitten, and bite into it— it will never be as fresh and delicious as it is just at this moment. In a few minutes, I load a crate with potatoes, beets, apples, more carrots, turnips, and a cabbage. Then, reluctantly, I lower the lid and shovel back the snow until the next time I run out of vegetables.

The kids are already back in the house, warming their frozen fingers and looking hungrily at the cookie tin. I start washing things for soup and hand them each a golden carrot. Their chatter subsides amidst the sound of crunching. My husband is another matter, and it's close on twelve, so I hastily assemble the rest of the soup and melt a few tablespoons of butter for garlic bread. Now's a good time, also, to think about what we'll be eating for the rest of the day. There's that great big cabbage; we could have cabbage rolls, stuffed with, hmmm, breadcrumbs fried with that bit of sausage I have out there in the freezer bucket? A little onion, a little sage? Right. Just then, the kids discover the apples, and I set them to chopping up a pile of them for an apple pie. By the time the soup is on the table, there's a pie cooking in my hot oven; by the time all the dishes are done, the oven temperature is sinking as I bank down the stove again, shove in a pan of cabbage rolls, and call it a day for the kitchen. We'll have rice with supper, and a cole slaw-sort of salad, but right now all I can think about is getting out in the dazzling sunshine.

This sort of morning may sound absurd to anybody who spends his or her winter looking out the window, but it's a demonstrable fact that people get a lot hungrier when they're actually outside in the cold of winter. They burn more calories, pump around more blood, and their entire bodies are subject to considerably more stress. They will not be satisfied with cups of coffee, tea, or soda, with just a piece of toast or a little salad for a meal. Lumberjacks and fishermen eat phenomenal amounts and, contrary to the cartoon image, are mainly a small and wiry lot.

To be comfortable, active, and healthy outside on a cold day, we need not look far for advice; history abounds with it. People living in cold climates all over the world have always recognized that food needs change with the seasons. The smorgasbords of Scandinavia, the rich meat dinners of England, the endless pastries and pies of Europe, and the deep-fried beancakes of the Orient all came about because people need more calories in the winter. But adding calories is not the only concern of a cold-climate cook. Vitamins, minerals, and cellular matter—fiber—have to be supplied in winter, when fresh foods are not as readily available as they are in the growing season. Fortunately, fruits and vegetables can be stored by many methods, ranging from the root cellar to the pickle barrel. Herbs and pot herbs, rich in minerals and vitamins, can be dried and stored.

Those who survived winters in past centuries ate large amounts of grains, threshed and winnowed on their own barn floors, ground at the mill and eaten along with the rest of what the Lord provided and the weather allowed. In much of twentieth-century America, this way of wintering has been all but forgotten. So much space is kept at such high temperatures that many people tend to dress unseasonably lightly in the winter. Between heated schools, stores, jobs and cars, we also get quite a lot less exercise during the colder months than people used to get. Living like this doesn't require many calories, as some of us have ruefully discovered. Many of our food habits—such as eating lightly during the day and relaxing with a big meal at night—add pounds and increase the likelihood of medical problems, and do little to keep our toes warm.

Perhaps the fuel crisis has arrived just in time to save us from ourselves, and restore energy and vigor to our lives. With less heat, we shall learn to cope with cold on more reasonable terms. We can wear warmer clothes, do more walking and exercising, and heat smaller areas for rest and relaxation. If these seem like sacrifices, we'll be able to make up for it in the food department. Eating more and richer food is a nice way to warm the heart.

I am not suggesting that we all return to the traditional groaning boards of northern farmhouses, however, just because we walk (instead of drive) to the supermarket, or keep the thermostat at 18°C instead of 24°C. Each individual has his or her own level of nutritional need, and although it is certain to increase with cold weather and added physical exercise, it's important for each of us to figure out what our own needs are and what each kind of food does for the body.

What Are Calories?

Calories used to confuse me when I read about them in high school. Most foods have them, yet they aren't a component of food in the sense that protein is a component of eggs, for example, or vitamin C of cabbage. On the other hand, it seems very important to know how many calories are in each meal, since an overabundance is stored as (often unsightly) fat. Eventually, I gave it up as too complex and found other things to do. The subject didn't come up again until a couple of years ago, when my family and I tried to get through winter on a Canadian homestead on a low-meat and low-fat diet. As we quickly discovered, calories are needed to provide energy to work and keep warm through a winter day; and without them, no amount of oranges and vitamin pills could keep us healthy and able.

A calorie is a unit of heat, measured as the amount of heat required to raise the temperature of 1 gram of water 1°C. In nutritional terms, it is a measurement of the amount of potential energy in any food. This is actual, usable energy, the kind your body cells burn, or oxidize, as you live and breathe and work and play and think and grow and fight diseases and keep warm. The more of these things you do, the more calories you need.

Calories come from three sources: fats, proteins, and carbohydrates. Fats have the most, at nine calories per gram; proteins and carbohydrates each have four calories a gram. But since these three different kinds of foods are absorbed in different ways, it's important to understand how they work in your body before you start figuring out what foods are best for you to eat, in what amounts, and when.

Protein: Since every body cell is made of protein, it is spoken of as an "essential building block." It's quite true that we can't live without it; on carbohydrates and fats alone, we'd starve. Large amounts of complete, ready-to-use protein are found in eggs, dairy products, and meats (including fish and poultry). Smaller amounts of less complete protein are found in seeds, nuts, grains, and legumes. By *complete,* I mean that the protein, which is made up of a couple of dozen amino acids, has the right amino acids in the right combination for your body to use.

Protein is absorbed gradually by the digestive tract, so the energy in it comes to you for hours after. It's important, therefore, to eat protein at the beginning of the day, before you begin needing energy.

Carbohydrates: This food group includes starches and sugars of all kinds: sucrose, from cane and maple sugar; glucose and fructose, from honey and fruits; lactose in milk; the starches in roots and grains. Foods high in carbohydrates tend to be digested very rapidly and converted almost immediately into blood sugar for a rapid increase of energy. This is sometimes helpful, when you need a quick, if short-lived burst of energy. On the other hand, it's important not to rely on carbohydrates for steady energy. Your energy level simply cannot be maintained for very long on carbohydrates alone. Be especially careful about foods with a lot of sugar in them. They can really throw off your system by making you think that you aren't hungry for the meal that, a few hours later, you'll wish you had eaten. Since all this is difficult for children to grasp, parents will have to play the heavy and limit their children's intake of sugar.

Fats: Fats have amost twice the amount of calories per gram that protein or carbohydrates have. Since you burn calories to keep warm, your need for rich, fatty foods goes up when the temperature drops. Fats are absorbed into the body even more slowly than protein. To help keep yourself warmer during a cold day, eat something with fat in it before you go out. This fat will be converted into blood sugar and will keep you relatively comfortable through the long hours outside.

If, however, you save your cravings for the end of the day and eat rich foods when you will be indoors, resting and sleeping for the next twelve hours, your body will have very little use for all that potential energy. It will be metabolized slowly and, if not called upon for energy, will be stored as body fat. This isn't so bad once in a while or on a small scale; but if it becomes a regular habit, you will soon find yourself overweight. Since extra pounds have to be carried around, heated, and taken care of, this puts additional stress on your body, particularly on your heart and circulatory system.

Why doesn't the stored fat get used up during the following day? Some of it does; we all make use of reserves from time to time. But the business of fat storage and retrieval is much more complex than simply burning calories available from that day's breakfast.

In order for the fat-retrieval system to function well, you must keep your body well supplied with other substances, including vitamin E and certain essential fatty acids, such as linoleic acid. These substances are not found in saturated fats, which include butter, margarine, hydrogenated lard or vegetable shortening, meat fats, or olive oil. There used to be a good supply in peanut butter, but nowadays it is usually "hydrogenated" (look at the label), making it just another saturated fat. Essential fatty acids and vitamin E are found only in unsaturated, cold-pressed natural oils and their sources—seeds, nuts, and whole grains.

You might wonder just how our forefathers survived with their pork-laden tables. Certainly there were no cold-pressed oils in pioneer days; why didn't they all become obese and drop dead from heart attacks? The answer is that, despite the high levels of saturated fats in their diets, they also ate whole grains, seeds, and nuts in volume. Refined flour and grain products are a comparatively recent invention, and have given rise to a new set of health problems.

Unsaturated oils also help prevent accumulations of cholesterol in your circulatory system. Cholesterol is a necessary part of your body's digestive and metabolic function; even if you don't eat any foods containing cholesterol, your body will manufacture its own, so there's no avoiding it. The only sensible thing to do is make sure you eat some foods containing substances that break down accumulations of cholesterol to keep your veins and arteries from clogging up. These substances are lecithin, cholin, and linoleic acid.

Is linoleic acid destroyed by heat? I don't think so, but cholin, lecithin, vitamin E, and other valuable nutrients are. Peanut, safflower, and sunflower oils are available at supermarkets at quite low prices, but the refining process that makes them so inexpensive involves both heat and chemical extraction. In other words, when you pay more for cold-pressed oils, you're getting more.

In the winter, when your fat intake tends to be higher than it is at other times, it is particularly important to include some unrefined cold-pressed oil in your diet every day. If you're unfamiliar with the tastes and qualities of different oils, march on down to the nearest natural food store and stock up on small jars of half a dozen varieties.

Store them in your refrigerator or cool pantry, and try them out on salads, where they will be of most benefit. Other uses include baking, sautéeing, sauce making, and general cookery; the heat will kill the lecithin, vitamin E, etc., but the oil is, nonetheless, unsaturated and contains the essential fatty acids that are so important in the winter. Recipes using these oils can be found on the following pages, and in dozens of natural-food cookbooks on the market today.

Vitamins and Minerals in Winter

Keeping yourself at an even 37°C (98.6°F) when the air is cold constitutes a major demand on your body. To keep you warm, your blood must move faster, your heart work harder, and your digestive system has a lot more food to deal with. Various glands and organs all over your body have to put in overtime to get it all straightened out. All these functions rely on a steady supply of vitamins and minerals to keep going. Going without one or another is like letting the timing of your car engine go awry. It will run, not well, but it may go on functioning at a low level of efficiency for a long time. Eventually, however, you wind up with engine failure.

Some foods are better sources of vitamins and minerals than others. They include:
- Fresh uncooked vegetables; lightly steamed vegetables, or vegetables in soups and stews
- Fresh and dried fruits; fruit juices
- Whole grains, whole-grain baked goods
- Fresh nuts, seeds, but butters
- Unrefined vegetable oils
- Molasses (unsulphured); blackstrap molasses
- Whole milk, skim milk, eggs, yogurt, cheese
- Liver, heart, and other organ meats

Whether or not you take vitamin pills, the main things is to plan your daily menus to include those things you know have real food value and avoid foods whose only virtue is that they taste seductive or are easily available. Healthful foods sometimes cost more, but there's more in them.

Stormy Days

When my brother and I were kids, in Bethany, Connecticut, snowstorms were absolutely our favorite things. The plough never came until all the other roads were done, because ours was the only dirt road in town. My father would go out and wrestle with chains and cinders and get stuck halfway out and stay home all day. It was a free day, unstructured, unplanned, with no dumb grownup plans except shoveling snow, which we kids regarded as pure sport, and a dim thought in the backs of our minds that we had, somehow, to keep warm. Our house had clearly been built by men who wore long woolen underwear all winter; and, anyway, on a day like that, who wanted to stay inside? Snow: we loved it, rolled in it, slid on it, piled it up high, made forts and mountains and roads for trucks, tanks, artillery—snowballs! We built long sweeping toboggan runs, to fly down faster than a leaf in a waterfall. And came inside, faces dripping, cheeks shining, aching inside with hunger. Whereupon my mother, bending over the big gas range, brought out pans upon pans of cookies: hearts and rabbits and stars. We gobbled them up with hardly a pause, and hungrily eyed the apple pie cooling on the windowsill, clearly slated for dessert, several thousand years away. Popcorn—we made loads of it, sitting by the fire in the living room, loading it with butter and salt. Thirsty, we prowled, searching the cupboards and freezer for cold fruit juice. And from the kitchen, miraculous smells would begin to filter through the house, something meaty and delicious: lamb stew! (My mother having providently set aside a stash of quick-thaw meat in the freezer, just in case.) Hot baked bread would issue forth next, the smell of it enough to knock you out. Tortured, we hung around the kitchen, unable to keep away from the heady aromas, the blazing warmth of the stove, the scurry of dinner preparations. Finally, after great clatterings of chairs and trips back to the kitchen to get this and that, we fell to. And those were the best meals, those dinners on our snowstorm days, watching the pure white flakes falling in the dusk outside the kitchen, watching as we sat warm inside, filling up on good lamb stew.

Large snowstorms, high winds, and cold snaps tend to come on suddenly and unexpectedly. In the "good old days" this didn't make very much difference to the kitchen; large amounts of food were stored over the entire winter anyway. Today, however, many households rely on a weekly shopping. If a big storm comes along just as you're running out of essentials, you'll really be stuck. So it's wise to purchase some extra supplies at the beginning of the hard-core months when you know storms are likely. Some can be dry-stored, some in a damp, cool place, and some you can keep in the freezer. Try to keep them separate from your everyday supplies and replace them if they get used up.

Winter Food Storage

Dry Goods: Canned meats, vegetables, fruits, and fruit juice can be stored almost anywhere.

Dry goods such as the following should be kept in tightly lidded glass jars to keep out damp and mold, since they may be stored longer than usual:

flour	salad oil
salt	cooking oil
baking powder	vinegar
dried yeast	coffee, tea
molasses, honey	whole-grain cereals
rice, bulgar	popping corn
dried beans	paper plates, cups, bowls
dried salted fish	dried milk (if you have kids, lots)
sproutable seeds	toilet paper
pet foods	paper towels

Vegetables: Certain vegetables keep well for periods ranging from two weeks to several months, depending on how cool you keep them. They include:

potatoes	cabbage
beets	kohlrabi
carrots	celeriac
turnips	

In a cool but very dry spot, you can store winter squashes and pumpkins.

Freezer: A great many perishable foods can be frozen: meats, butter, vegetables, fruits, fruit juices, and baked goods. If your freezer is quite small, think about what will be most important if you get snowed in for a few days. Small freezers often don't keep food as cold as big ones do (constant −18°C [0°F] is optimum) so you may need to use up and replace your supplies once a month or so.

Refrigerator or Cold Pantry: If you don't have much freezer space or for certain nonfreezable but somewhat perishable items, try this: Using a shoebox or plastic refrigerator storage bin to ensure that no one confuses it with the everyday food, set aside a few items like eggs, butter, cheese, oranges, apples, cured sausage. Each week, take them out and replace with fresh. If you don't use a lot of these things, store only a little; but if your family is large, an extra dozen eggs or oranges can be a real godsend.

Cooking without Electricity: If you're dependent on electricity for cooking, you should certainly invest in an emergency heat source in case of electrical blackout. A white gas-fueled camp stove is a good alternative. These stoves come in a variety of shapes and sizes, ranging from a pocket-sized single burner to a larger two-burner unit, suitable for cooking regular meals. Make sure you have a can of gas (clearly marked), a funnel, and a little practice in using the stove before you're faced with all those problems on a dim and stormy night. You'll also need light to cook by; candles are pretty, but frustrating when you're trying to keep track of the spatula. Make sure you have at least one good kerosene lamp (tried and true) and a supply of kerosene, also clearly labeled. Whatever you do, don't get kerosene and white gas mixed up or you'll have a fire. In case of that, remember that you can't put out this kind of fire with water. A foam-type fire extinguisher or a big box of baking powder should always be on hand. Cooking is a little different on camp stoves. You should think of dishes that require few pots, such as stew, goulash, eggs and potatoes, or rice and stir-fried vegetables.

Water: With all these preparations, don't forget water. When a storm warning comes over the air, fill a clean five-gallon lidded container with water for drinking and as many buckets as you can, or the bathtub, with water for washing. Wash all dishes and pans as soon as possible after using them.

In General: Try to think of everything before the need arises. For example, don't let yourself run low on medicine, special dietary needs such as baby foods, or livestock feed. (We keep an extra sack of each kind of feed on our farm throughout the winter.) If you do get caught, don't panic and try to get "through" to an outside world that may not even exist, what with everybody else being at home, drying socks and reading about wood heat. If you're short of something, make do with something else until the storm is over and the world is ploughed out again. Relax, enjoy yourself, and cook up a storm.

Breakfast
Hot Spiced Applesauce*
Fried Eggs
Home Fried Potatoes
Scones*
Milk, Butter, Jam
Coffee or Tea

Lunch
Minestrone Soup*
Whole-Wheat Bread, Butter
Cheese

Afternoon Snack
Apple Pie*
Milk, Hot Herbal Tea

Supper
Stuffed Cabbage Rolls
Brown Rice
Salad

*Asterisks mean you will find my recipe
for this dish in the pages that follow.

Hot Spiced Applesauce
makes about 2 quarts

Applesauce is so easy to make—and is such a
good way to use up bruised and mealy
apples—that you may never use the store-
bought kind again.

Quarter and core:
   3 pounds of apples
If you have a sieve or a food mill, you don't
need to peel them; your applesauce will be a
lovely pink if you leave the peels on.
Pile the apples into a large, heavy pot. Add:
   1/4 cup water
   1-2 tablespoons of brown sugar (depending on whether your apples are tart or
   sweet)
   Juice of 1/2 lemon
   A 5 cm (2") stick of cinnamon
   3 whole cloves
   1/4 teaspoon of mace or allspice
Cover the pot and set over a low flame.
Check again in about 15 minutes and stir up
the apples from the bottom. As soon as they
begin to soften and lose their shape, remove
the cover and cook until they are mush. Fish
out the cinnamon stick and the cloves.
(Don't worry if you can't find them; they'll
be stopped by the sieve.)
   Pass through a food mill or push through
a sieve with a wooden spoon. Taste to see if
it is spicy and sweet enough, and adjust
seasonings to suit your fancy.
   The applesauce can be put into canning
jars and processed, or stored for a while in
the fridge. (Remember, you didn't add any
BHT, but then your family won't let it stay
around long enough to spoil.) Reheat in a
saucepan or serve chilled.

Scones
makes 8 scones

In a large mixing bowl, combine:
   2 cups flour
   2 teaspoons baking powder
   1/2 teaspoon salt
Mix in, using a fork and stirring well:
   1/3 cup corn oil
Add gradually, stirring and mixing thoroughly:
   about 1/3 cup milk
Don't allow the dough to get sticky; stop
adding milk when you can shape the mixture
into a sort of oversized softball. Divide the

dough into two hunks, and roll out each one
about 1.3 cm (1/2") thick. Cut into wedges
(called *farls* in Scotland); you should be
able to get four out of each piece of dough.
Prick with a fork. Bake on top of the stove
in a preheated, well-seasoned skillet, over
moderate heat, for 5 minutes; turn, and bake
5 minutes longer. These can also be baked
on the top surface of a wood stove, provided the top is kept rust-free by rubbing
in vegetable oil instead of stove blacking.

Variations
For flour, you may use:
all whole-wheat pastry flour;
half whole-wheat bread flour and half
white unbleached *or* oat flour; all oat flour;
three-quarters whole-wheat pastry flour
and one-quarter roasted, ground sunflower
or sesame seeds

Minestrone Soup
serves 6

From the Italian, "to administer, take care
of all one's needs"—this soup does all of
that and more on a winter day. It's a meal
in itself. You may make it with stock, as I
did in this version, or cook some meaty
bones along with the broth, then remove
them and chop up the meat at the end. If
you like, you may leave out the meat and
the bones altogether. If you do not use
stock, however, make up the liquid difference with 2 more cups of water.

Wash, pick over, and soak overnight in 1
quart water:
   1 cup pea beans (pinto, Roman, kidney,
   navy, and others will do too)
Next morning, drain the beans and discard
the soaking water (which is filled with gas-
producing enzymes).
In a heavy 5-quart pot, heat:
   1 tablespoon vegetable oil
Add and sauté briefly:
   1 medium onion, chopped
   2 stalks celery, chopped
   1 clove garlic, minced
Then add:
   2½ quarts water
   2 cups beef stock
   1½ teaspoon salt
   soaked beans
   1/2 cup barley
   1 bay leaf
   1 teaspoon basil
   1/2 teaspoon oregano
   1/2 teaspoon thyme
Bring to a boil, cover, and simmer 3 hours or
until beans and barley are soft. Add:
   1 cup cooked or canned tomatoes,
   chopped
   1/2 cup uncooked egg noodles
   1/2 cup fresh or frozen vegetables
   (choose peas, green beans, corn, potatoes, turnips, squash, or cabbage), diced
   in 1.3 cm (1/2") squares
Optional:
   1/2 to 1 cup cooked meat, chopped
   2 tablespoons fresh parsley, finely
   minced
Cook rapidly for 10 minutes, until noodles
and vegetables are tender. Adjust seasoning;
grate in a little fresh pepper. This soup will
keep refrigerated for days, but be careful
not to overcook the noodles when you reheat it.
   Serve with garlic bread and pass around a
big bowl of freshly grated hard cheese, like
Parmesan or Romano.

Scone farls

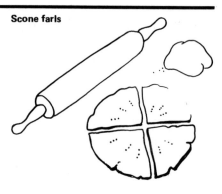

Apple Pie

To make a really good piecrust, your shortening must be cold, your flour must be light, and your oven must be very hot. There are various schools of thought regarding ingredients. Some feel that a light, crisp, flaky texture can be achieved only with white flour and lard. Others prefer to work with ingredients that are more nutritionally sound —whole-wheat or other whole-grain flour and oil. These crusts are nutty and delicious, but practically impossible to roll out; they are crumbly rather than flaky. I use them in many of my pies, but in the case of apple pie, I think it makes more sense to use a roll-out piecrust, because the apples need a fairly tight cover to steam properly. Hence, I devised this recipe, using whole-wheat pastry flour, which is made from a softer variety of wheat than all-purpose whole-wheat flour.

Mix in a chilled bowl:
    1½ cups whole-wheat pastry flour
    2/3 cups cold butter or home-rendered lard
    1/2 teaspoon salt
    1/2 cup wheat germ
Mash these together with a fork until the butter or lard is is in tiny lumps, and the texture of the mixture is like coarse cornmeal. Slowly add about 1/4 cup cold water in sprinkles, until the dough holds together. Depending on the dampness of the day and the age of the flour, you may need more or less water. Gauge it by how well it holds together, not by how much water you have used. Refrigerate, covered, 1 hour or longer.
Preheat oven to 400°F (204°C)* and chop up:
    Apples to yield 6-7 cups
Mix these up with:
    1 tablespoon cornstarch or flour
    1 teaspoon cinnamon
    1 teaspoon allspice
    1/4 teaspoon nutmeg
    1/3 cup yogurt
    2/3 cup brown sugar
    1 teaspoon lemon juice
Mix well to coat all apples evenly.
    Roll out piecrust, using two-thirds of it for the bottom crust. Whole-wheat pastry flour crusts are easy to roll out, but if you are using hard whole-wheat flour, you may find it easier to roll out crusts on a piece of floured canvas (see illustration). Line buttered pie plate with crust, trim, and fill with apples.

Dot the top of the apple mixture with:
    2 tablespoons butter

Roll out top crust. Using your finger, moisten the rim of the lower crust and place the top crust over the pie. Trim it around, then fasten the two crusts together either by pinching at intervals, or with a traditional fork pattern. Prick the top a few times with a fork.
    Bake at 400°F (204°C) for 15 minutes; then lower heat and cook at 350°F (177°C) for another half hour.
    Serve warm or cooled to room temperature. A slice of cheddar cheese is an irresistible accompaniment; those who will go outside to work or play after the snack should feel free to indulge.

*Throughout this chapter, Celsius equivalents are given in parentheses since, despite the conversion to metric temperature measurement, most of us still have Farenheit scales on our oven dials.

Turning a pie crust on a canvas or cloth

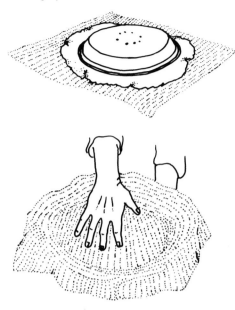

Keeping Children Warm

Children, in general, manage to stay warmer in the winter than adults. Being shorter of limb, their circulation is better. Most younger children also have outer layers of natural fat, which help insulate them against cold. On the other hand, and partly for these reasons, kids tend to spend a lot more time outside than adults do. So, in the end, they wind up needing just about the same amount of extra calories in the winter as anybody else.

Children, however, have eating habits somewhat different from those of adults. Because they are growing, their needs for protein, calcium and other minerals, and vitamins are higher than you might expect. On the other hand, their stomachs are small. Too often, they run out of fuel before the next mealtime and, naturally, they hunt for snacks. If the snacks are high in sugar, and eaten shortly before a meal, the child won't want to eat much of anything until just about the time you finish the dishes.

There are only two sane things you can do in this situation. One is to serve breakfasts and lunches with enough nourishment and variety to hold them through to the next meal. The other is to back that up by leaving only healthful things to munch on around the kitchen. A plate of "kid salad" (p. 99) makes a good hors d'oeuvre. For between meals: whole-wheat rolls are less trouble than bread, which has to be sliced. Blackstrap molasses or regular, unsulphured molasses is good to spread on it, and doesn't promote cavities. In the refrigerator, establish a shelf or box containing "free" munchables: cheese, yogurt, fruit, whatever you see fit to dispense. In jars in the cupboard, you can store nuts and dried fruits, nut butters, healthy-type cookies. If your children depend on snacks, make sure they have good food for snacks; and if they snack on good food, you need not worry too much about whether or not they'll eat their dinner.

Menus for Keeping Children Warm

Breakfast
Hot Cereal*
Dried Fruit
Milk

Lunch
Scotch Broth*
Sesame Biscuits*
Milk of Juice

Afternoon Snack
Baked Apple
Oatmeal Spice Cookies*
Milk

Supper
Lamb Stew*
Kid Salad*
Milk or Juice

Hot Cereal
Hot cereal is a wonderful, filling, quick, and easy way to start a busy day. Keep several kinds in stock and don't be afraid to try the ones that take 10 or 15 minutes—they always taste better than the ready-in-a-minute sort.

As soon as you get up, pad into the kitchen and set a kettle on to boil, and measure cereal and salt into the pot. Set out some bowls, spoons, and whatever you have to put on the cereal. By this time, the water will be boiling; stir it into the cereal, cover, and set to simmer over low heat so it won't boil over.

The success of hot cereal as a breakfast will probably depend mostly on the variety of toppings you offer. Some people like nuts and coconut; others are crazy over raisins. My grandmother taught me to enjoy butter instead of sweets on my cereal, and I eat it that way to this day.

Here are some lovely things to set out:
broken walnut meats
unsweetened coconut shreds
dried fruit, including currants, dates, apricots, raisins, pears, peaches, pineapple
tamari soy sauce
brewer's yeast
brown sugar
honey or maple syrup
butter
milk or cream
yogurt
wheat germ

Scotch Broth
serves 6-8

In a pinch, any bones will do for a good soup; but there's something special about the marriage of lamb and barley. This is a great big pot of soup and it will keep for days, refrigerated; it gets better as time goes on.

Soak overnight:
1/2 cup white navy beans in 1 quart water
In the morning, drain and rinse the beans. Discard soaking water.

In a large pot, heat:
3 tablespoons corn oil
Sauté:
1 medium onion, chopped
1 carrot, chopped
Add and bring to a boil:
2 quarts water
1 teaspoon salt
1 cup pot barley
the soaked beans
1 teaspoon thyme
1 bay leaf
lamb or mutton bones
The best bones are those big thick ones you get in a roast, but any bones add flavor and calcium. They can be precooked or raw. If you have a little chewable meat to add, chop it finely and wait until the soup is almost done to add it. Simmer the soup, covered, until beans and barley are done, about 2 hours. Remove bones and add:
1 carrot, chopped
1/2 cup chopped lamb or mutton (cooked or raw)
1 cup frozen peas
Simmer 10 minutes and serve.

Sesame Biscuits

Two dozen 3.8 cm (1½'') rounds; a dozen 7.5 cm (3'') rounds;

A rich, flavorful biscuit, made from whole-wheat flour and sesame seeds.

Preheat oven to 450°F (230°C)
Roast carefully until golden, over moderate heat in a frying pan:
    1/2 cup sesame seeds
Take care not to let them burn.

Grind or pound these until they are mostly flour-like. You can use a grain or coffee grinder or Japanese surbachi, or a mortar and pestle.
Sift together:
    2 cups whole-wheat pastry flour
    1½ teaspoon baking powder
    1/2 teaspoon salt
Add:
    ground sesame seeds
    1/2 cup cold shortening (lard or butter)
Mash with a fork until the lumps of shortening are small, but not completely gone.

Make a well in the center and add:
    3/4 cup cold milk
Beat well; turn the dough out on a floured board and mix it slightly with your fingers until it sticks together in a lump and can be rolled out.
    Roll out 1.3 cm (1/2'') or thicker, as desired. (The buscuits will not puff up as they cook.) Cut into shapes with a cookie cutter or a drinking glass. Bake on a lightly oiled cookie sheet, for about 10 minutes.
    If you have time, these biscuits look very fancy when brushed with beaten egg yolk and sprinkled with sesame seeds before baking.

Oatmeal Spice Cookies

30 cookies

A cookie recipe without sugar is worth a million Oreos when it comes to saving teeth, so says my dentist.

In a large bowl, mix:
    1/4 cup vegetable oil
    1/3 cup molasses
    1 egg
When beaten, add:
    2 cups rolled oats
    1/2 cup dried milk
    1 cup whole-wheat flour
    1/2 teaspoon cinnamon
    1/4 teaspoon powdered ginger
    a pinch of ground cloves
    a grating of nutmeg
Top-stir the dry ingredients a bit and then add:
    1/2 cup currants
    1/2 cup chopped dates or figs
Mix the fruit with the dry ingredients, then stir the whole mixture together. With damp palms, roll 1½-tablespoon balls of dough, and flatten them on a lightly oiled cookie sheet.

Bake 10 minutes at 350°F (177°C).

Lamb Stew

4 or 5 servings

Dredge in flour:
    .9 kilograms (2 pounds) cubed stewing lamb
In a large heavy pot, heat:
    3 tablespoons vegetable oil
Brown pieces of meat lightly on all sides, and remove from pot.

Sauté in pot:
    1 onion, well chopped
    1 carrot, chopped
    1 clove garlic, minced
Return lamb to pot, along with:
    4 cups water
    3 sprigs fresh parsley
    1 bay leaf
    1 teaspoon thyme
Cover and simmer for an hour or so. Then add:
    6 medium sized potatoes, scrubbed and quartered
    4 carrots, scrubbed and quartered
Simmer 1/2 hour or until potatoes are almost done.

Add:
    1 cup frozen peas
    1 teaspoon salt
Simmer 10 minutes or until peas are just done. Serve hot. This stew is even better the second day, as long as the peas don't get overcooked.

Kid Salad

From the time they start teething until they're fully grown, kids seem to enjoy fresh, raw vegetables better than cooked ones. Some of these are probably familiar to you; others may be unexpected, but they are worth trying. Fill a dinner dish or tray with an assortment of:
    strips of green or red pepper
    slices or sticks of fresh raw beets, kohlrabi, celeriac, turnips, carrots, celery
    alfalfa and mung bean sprouts (for growing instructions, see p. 104)
    radishes
    broccoli and cauliflower flowerets
    cabbage hearts, Brussels sprouts
You may accompany this with a small dish of olive oil or mayonnaise as a dip.

Winter Dieting

Winter is a great time to lose weight, because you burn calories just by being outside in the cold. Moreover, you can't stay out long without engaging in some form of exercise (to keep your circulation going), which also contributes to healthy weight loss as well as building up smooth, elastic muscles and skin. The renewed interest in cross-country skiing, outdoor skating, walking, and other winter outdoor activities is a truly healthy trend.

To enjoy the full benefits of such exercise, however, you must be careful to support your body with adequate nutrition. When people diet, they all too frequently get carried away by calorie counting and tend to forget that there are other aspects to food besides their caloric value. There are a great many substances in a normal diet that are essential—not just recommended, but *essential*—to your health and well-being. You may be able to get along without them for a while, but you wouldn't be happy about the deprivation if you knew what was happening inside you to compensate for the nutritional imbalance. I don't think it's healthy to go without whole grains, fresh fruits and vegetables, unrefined vegetable oils, or an adequate level of protein. Nor do I believe that vitamin pills can provide enough sustenance in the right proportions for the average person. Your own appetite is the best gauge of what you need, as long as it isn't distorted by junk foods or drugs. The real answer to staying slim is to eat foods that support health and to get enough exercise the year 'round.

Another good reason for a sensible approach to weight loss is that a crash diet doesn't really change the eating habits that are responsible for the excess fat in the first place. People who gain unnecessary pounds do so because they eat more calories than they can use—usually, at a time of day when they don't need them. By getting started on better eating habits, and learning which foods you need when and why, you can avoid regaining weight after the view from your mirror is acceptable again. Weight loss may not be so dramatic with gradual dieting, but it lasts longer—and it won't land you in the hospital.

To begin, a high-energy breakfast is essential. It may sound crazy at first, but if you're serious about losing weight, don't go hungry in the morning. Get up early and relax with a good meal, which should include a high level of protein as well as some carbohydrates and fats. For example, fry your eggs in a little oil instead of butter. Eat whole grains instead of refined white bread; you need the roughage to digest your food easily. Since your food intake will be somewhat smaller and more concentrated than usual, this is especially important. Whole-wheat toast, muffins, whole-grain cereals, or even a couple of teaspoons each of bran and wheat germ will do the trick. Finally, include some real fruit in your breakfast. Don't bother with those instant crystals or "fruit-flavored" drinks. They have too much sugar and not much else. Frozen fruit juices are fine, as long as they are freshly made, and fruit compote is terrific for breakfast: stew up a couple of prunes, a dried apricot, and a slice of lemon per serving the night before, and refrigerate until breakfast time.

The next step to losing weight is to move dinner up to lunchtime. This means that you will be getting the full benefit of the meal when you most need it—during the day. (Also, if you should happen to eat more calories than you need, you'll have the chance to work them off during the afternoon.) For this meal choose a portion from each group:

1 *High-Protein, Low-Fat Foods:*

Broiled steak	Tofu (bean curd)
Broiled chicken	Cottage cheese
Broiled calf liver*	Soyburgers
Broiled lean fish, such as	
haddock, halibut, cod,	
flounder, sole, bass	

2 *Low-Starch Vegetables*

Broccoli	Green or yellow beans
Cabbage	Peas
Brussels sprouts	

3 If you're a heavy eater, or just having a heavy day, also serve:

Beets	Winter squash
Carrots	Parsnips

4 Always have a salad made out of fresh, uncooked vegetables, in addition to the cooked vegetables:

Lettuce	Endive
Tomatoes	Parsley
Mushrooms	Grated carrots
Green peppers	Radishes
Sprouted seeds or beans	

Dress your salad lightly with unrefined vegetable oil (safflower, peanut, corn) and one or more of the following:

Vinegar	Tamari soy sauce
Lemon juice	Yogurt
Onion juice	Mayonnaise

*Much tastier than beef liver; if you've never liked liver, it's worth a try.

Such a meal will be full of nutrition and low in fat, particularly animal fats. Don't cook with lard or bacon fat or butter, if you can help it; substitute vegetable oil and use it sparingly. Without animal fats (which are commonly supplied in fried foods), pastries, sauces, gravies, and rich meats, your body will be forced to use up some of its stored fat, especially if you spend any part of the day exercising outdoors. The small amounts of vegetable oils are needed so that your system can unlock stored fats and liquids—in other words, so you will lose weight.

By mid-afternoon, you may need a snack. This is a low point in the day for many people, and it's important not to let your hunger build up to a degree that you find it difficult to stick to dieting. Choose something that suits the day. For example, if you've been working in an office all day and plan to walk home, a piece of fruit, low-fat whole-grain crackers, and cheese might be just the thing. If you're sitting outside school in a car waiting for your kids to pile in, take along some yogurt or cottage cheese. If you can't stand diets without secret sinning, have a couple of oatmeal spice cookies or a small piece of pumpkin pie and a glass of milk before you go for a jog. Whatever you eat, though, choose it because it has nutritional value above and beyond its calories.

At the end of a dieting day, take it easy. This is the time when you really have to watch what you eat, especially in the winter, when evenings are long and mostly spent indoors. You won't need many calories, so don't consume them. A light supper is best—soup and salad and maybe some crackers or toast, a little cheese, a glass of milk, some fruit if you haven't had any yet. If you'd rather, you can space these things out, having a soup supper and a fruit and cheese snack later.

Having a very light dinner is sometimes difficult. For many people, the evening meal is the only shared meal of the day, or the only time that's feasible for entertaining guests. In that case, serve something low in fat—say, fish filets—along with an interesting vegetable and a glorious salad. You could add potatoes, rice, or noodles, but eat little of them yourself.

Finally, keep foods around your house that you can really depend on as sources of lots of important nutrition, and experiment with using them in different ways. Quite often, a craving for a particular food is actually based in a body need. For example, you crave ice cream because you need calcium; drink some whole milk instead, or have yogurt and honey. Here are some suggested sources of other "extraordinary" nutritional needs:

For extra iron:
> Blackstrap and unsulphured molasses
> Dried fruit (raisins, currants, prunes, apricots)

For extra calcium:
> Milk
> Cheese
> Yogurt
> Bone broth

For extra B vitamins:
> Liver, organ meats
> Debittered brewer's yeast
> Soy products
> Wheat germ

For extra vitamin E:
> Wheat germ
> Unhydrogenated peanut butter, nuts, seeds
> Whole grains and whole-grain products

For extra vitamin C:
> Oranges, lemons, grapefruit
> Fresh cabbage, broccoli, Brussels sprouts
> Fresh and canned tomatoes

For extra vitamin A:
> Dried apricots
> Carrots
> Egg yolks
> Liver
> Cod-liver oil

For extra vitamin D:
> Cod-liver oil

These substances are found, in small amounts, in other foods, but since you are eating less food than usual, you may need to supplement your diet by eating foods in which nutrition is more concentrated.

Finally, find time to go over your diet once a week, and make changes as needed before you do your next week's planning and shopping. As the weather gets colder, you may find you need more protein-rich foods, even more fats; but it doesn't stay cold forever, so a regular reevaluation is important. When your winter diet fits your particular needs, it's much more likely that you'll stick with it; you'll shed unnecessary pounds, and keep your figure in line all year 'round.

Breakfast

1 egg, soft-boiled or poached on whole-wheat toast
1 glass skim or whole milk or 1 slice cheese
1 glass fresh or frozen fruit juice or All-in-One Breakfast Drink*

Main Meal (Lunch or Dinner)

1 or 2 filets of low-fat fish, such as haddock, sole, cod, brushed with oil and broiled
steamed broccoli or Brussels sprouts
Tzimmes*
Crisp Winter Salad*

Afternoon Snack

2 Whole-Wheat Muffins*
or 1 cup yogurt, lightly sweetened with honey and sprinkled with a tablespoon of wheat germ

Small Meal (Lunch or Supper)

quick chicken noodle soup
rye crackers, Swiss cheese
fresh fruit

All-in-One Breakfast Drink

If for some reason you find it impossible to get a balanced breakfast, try this:

In a blender, or with an eggbeater, combine:
    1 cup milk
    1 egg yolk
    1/4 cup dried skim milk powder
    1 teaspoon brewer's yeast
    1 tablespoon safflower oil
    1 tablespoon carob powder or 1/4 cup orange juice
If you use brewer's yeast every day, you can gradually increase the amount to 2 tablespoons, which is much better for you, but likely to cause gas problems if you're not used to it.

Whole-Wheat Muffins
makes 15

If you haven't been getting much in the way of whole grains, this is a tasty way to add them to your diet. These muffins offer lots of vitamins, minerals, and roughage, as well as good eating.

Preheat the oven to 400°F (205°C)
Combine in a mixing bowl:
    2 cups whole-wheat flour
    2 teaspoons baking powder
    1/2 teaspoon salt
    2 tablespoons wheat germ (optional)
Mix together, then add:
    1 egg
    1½ cups skim milk
    1/4 cup molasses
    1/4 cup corn oil or safflower oil
Beat to combine in a few strokes and pour into oiled muffin cups.

Bake for 20 minutes.

Tzimmes
serves 6

This is an old Eastern European dish, dear to the heart of many a Jewish housewife. Its sweet-tart flavor goes well with liver or fish.

Peel and slice:
    1.3 cm (1/2") thick, one medium-size winter squash, such as acorn, butternut, or hubbard
    .6 cm (1/4") thick, 4 carrots

Pile the carrot slices in a vegetable steamer and lay the squash pieces on top. Steam until tender, about 20 minutes (test with a fork).

Meanwhile, oil a two-quart casserole or pot that has a good tight lid.

Chop up:
    4 tart apples
When the vegetables are done, arrange squash, carrots, and apples in layers (more or less) and top with a mixture of:
    1/4 cup vegetable oil
    1/3 cup honey or molasses
    1/4 teaspoon salt
    grated lemon rind (only if you use honey)
    a squeeze of lemon juice
Cover the casserole tightly, and bake at 325°F (165°C) for half an hour.
Remove the lid for the last 10 minutes to give the dish a slight glaze.

Crisp Winter Salads

Whether you're dieting or not, salads are wonderful. In the winter, though, the available ingredients aren't very interesting: lettuces tend to be bitter, tomatoes mushy, and both are ridiculously expensive. Growing sprouts is a handy solution. They are always fresh, tender, and delicious. The price is unbeatable too. Supplement them with vegetables that keep well in the colder months. Grate carrots, beets, cabbages, kohlrabi, celeriac; chop in some raw broccoli, cauliflower, celery, and Brussels sprouts. The main thing is to use vegetables that are fresh and crisp, kept in very cold storage (just above freezing) and damp (not wet, but humid) right up to the moment you prepare and eat them. I keep most of these in the root cellar. If you're buying your salad vegetables, shop for them on the day they arrive in the store and store them in the refrigerator as soon as you get home.

Savoy-Sprout Salad
serves 4

Grate and mix:
    1 cup savoy or ballhead cabbage
    2 cups carrots
To grate cabbages, first quarter them, cutting through the stem so that they hold together as you grate them.
Add:
    1 cup alfalfa sprouts
    1/2 cup lentil sprouts
    1/2 cup dried currants or chopped figs
Toss with:
    2 tablespoons mild-tasting unrefined vegetable oil
    1 tablespoon freshly squeezed lemon juice
Then toss in:
    1/2 cup mayonnaise
Serve at once or chill until mealtime.

Purple Sprout Salad

Grate and mix:
    2 cups red cabbage
    1 cup beets
    1/4 cup onion
Add:
    1 cup alfalfa sprouts
    1/2 cup mung bean sprouts
    2 tablespoons fresh chopped parsley
Toss with:
    3 tablespoons mild-tasting unrefined vegetable oil
    2 tablespoons red wine vinegar or cider vinegar

Then toss in:
    1/2 to 1 cup yogurt
Serve at once or chill, covered, until serving time.

Sprouting Seeds

Sprouts are a wonderful source of fresh, crisp, low-calorie salad material. They taste great, and they're loaded with vitamins and proteins. They're also great fun to grow. As you get the hang of it, you can branch out and try all sorts of seeds, singly or in combinations. The method is simple: You put the seeds in a jar, soak them in water, drain them, and then keep them warm, damp, and dark for four to seven days, rinsing them every so often. In our house, we keep several jars of different varieties of sprouts going all the time. I start a batch every two or three days, so there's always a new jar coming along. You don't need any fancy equipment to sprout seeds.

Three sprout jars

Canning Jars: Use a 1- or 2-liter (1- or 2-quart) canning jar with screwtop ring. Cut a 15 cm (6") square of nonrusting window screen; place it over the top of the jar and screw on the ring. The only problem with canning jars is that the neck is a little narrower than the base of the jar, so you have to be careful to drain out all excess moisture when you rinse seeds, especially small ones, like alfalfa seeds.

Straight-Sided Jars: Use a 1- or 2-liter (1- or 2-quart) jar. If a canning ring doesn't fit on it, cut loosely woven cloth (cheesecloth is ideal) to overlap the top by at least 7.5 cm (3"). Use big rubber bands to hold on the cloth.

Clay Pots: Buy a new unglazed flowerpot. Fit a circle of nonrusting window screen in the bottom, and cut a piece of cloth to wad in the top. Soak the pot in water for several days before using. The advantage of this method is that it excludes light and keeps the seeds moist and well drained at all times. The only drawback is that if a batch of seeds happens to go bad (they sometimes ferment) in this container, you will have to boil it out for 10 minutes in a big pot of water to kill any bacteria lurking in the pores of the clay.

Measuring Seeds: Sprouts expand tremendously, so don't put too many seeds in a jar. For a quart jar, use:
  2 tablespoons alfalfa seeds, or
  1/4 cup mung beans, lentils, soybeans or other legumes, or
  3 tablespoons radish, mustard, cress seeds
Before you use them, pour your measured seeds out on a dinner plate and remove any that are broken, shriveled, or moldy. This is especially important in the case of legumes.

Growing Sprouts
First of all, rinse your seeds. Put them in the container and fill with cool water, then let it drain out. Next, fill the jar with tepid water (never use hot!). If you're using a flowerpot, set pot and seeds in a bowl and fill with water within 2.5 cm (1") of the top of the pot. Soak seeds four hours. Then drain, and rinse again with cool water.

Place your container in a dark, warm place for the seeds to grow. The temperature should be about 20°C (70°F), so don't try putting them in a wood-stove warming oven! If you're using jars, lay them on their sides, tilted slightly so any excess moisture can drain out onto a plate or dishcloth. If the jar is in the light, cover it with a cloth. If using flowerpots, set a saucer underneath to catch the drips.

Every six hours or so, you should rinse the seeds by filling the container with cool water and letting it drain out. Six hours is optimum; but nobody gets up in the middle of the night to water sprouts. Three times a day is fine, and I have even gotten away with morning and night rinsings, although my seeds grew somewhat slowly.

Once in a while, you experience a sprouting disaster. The seeds ferment and give off a bad smell. This is usually caused by poor drainage and not enough rinsing. Chuck out the seeds and wash the container very well; if it's a pot, boil it.

In general, you can expect sprouts to appear in a couple of days. After that the length they grow to varies with the type of seed. Alfalfa sprouts can become 2.5 cm - 5 cm (1" - 2") long, making a lovely, delicate tangle from just a few seeds. Mung beans make a long, fat sprout, often used in cooking Chinese mixed vegetable dishes. Lentils, too, are good to cook with, and have a smaller seed and more delicate flavor than mung beans. Other seeds used by adventurous sprouters include:

lima beans	wheat
black beans	rye
chick peas	corn
Windsor beans	barley
adzuki beans	cress
soybeans	mustard
kidney beans	radish
oats	flax

Types of sprouts

lentils   mung beans

wheat   alfalfa

All sprouts are more nutritious raw, since some of their vitamins are destroyed by heat. In my kitchen, however, I tend to use the legumes more in cooking, since the beans are larger and seem to taste better sautéed in a little vegetable oil. Sprouting is a wonderful way to use beans; the continuous rinsing reduces their gas-producing qualities and makes them more digestible than ordinary cooked beans.

The grains take longer to get going than other sprouts, and should be used when the sprout is still very short. They are nice and chewy, can be used in both salads and cooking, or added to bread dough. Not all grains will sprout. This is because whole grains are frequently steam-cleaned before they are packaged. This in effect kills the grain. Look for grains with the word **sprouting**, or ask the purveyor.

Cress, mustard, radish, and alfalfa can be "greened" after they are fully grown. When you see pale leaves forming on the ends of the tendrils, set the jar on the windowsill for a day (don't forget to keep rinsing, though). Greened sprouts are a little chewier and have a more biting flavor than sprouts that have not been exposed to sunlight, but they also have more vitamin C and look inviting on the table.

Sources of Seeds
Health-food stores and natural-food dealers sell sproutable seeds. Sometimes you can get them from seed catalogs, but make sure they haven't been sprayed or chemically treated. For best results, they should be used within a year of harvest, and kept in a cool, dry place—in jars, tightly lidded.

Special Meals

Everybody likes to do a little fancy cookery once in a while, and it can be an especially nice way to spend time in the winter. Personally, I find blizzards particularly inspiring. When the winds are really howling and the snow piles deep around my door, you can always count on a feast at our table. Other people really get into it for special occasions; some of the most exotic dinners I've ever had were prepared by a friend who always tries something totally new when she is having a dinner party.

Whatever your reasons, cooking fancy can be great fun, and hunting through cookbooks and magazines in search of new recipes is only part of it. The real test of kitchen mastery lies in how you relate to the food itself. Each ingredient has its own essence. An egg or a potato can be young or old, one of many shapes and sizes, each with different possibilities. Combining ingredients into dishes, courses, and meals is a creative activity.

Getting on Top of a Cold

Almost everybody is exposed to cold and flu germs at one time or another during the winter. They're everywhere, especially in closed, heated buildings. They float around waiting for a warm, damp place—like your nostrils or throat—to settle down. Once they're there, they hatch out of their dormant state and start dividing and multiplying. A tickle in your nose turns into an infected sinus, spreads down your throat, and there you are: you have a cold.

Some people never seem to get colds. They're exposed to just as many viruses as anybody else, but their "natural resistance" is higher. When germs land in their nose and throat, their body's white blood cells gobble up the intruders and halt the progress of the cold before it gets started.

Suppose you are a chronic cold victim and you would like to try to increase your resistance. How can you do it? First of all, pay attention to the mild symptoms of a cold before they get serious. It's much easier for your body to get rid of the infection right at the beginning, before it spreads. Declare yourself sick, cheerfully and firmly, and head for the kitchen. Your body will need more protein, more liquid, and plenty of vitamins—not only C, but all kinds.

Stay away from rich foods, bulky foods, foods with a lot of fat; these are too hard for you to digest when your body is busy mending fences. You should be resting, so all your energy can go into getting better. Have lots of low-fat soups and fruit juices, with a few whole-grain crackers to munch on. If you like yogurt, that's great for a cold, partly because it's so digestible, partly because it replaces friendly intestinal bacteria that are often destroyed by the cold or by antibiotics. If you don't like yogurt, have a little cottage cheese during the day. Snack often, and keep a good collection of fresh fruit at your elbow. Have your meal in the middle of the day, and choose something light—fish, chicken, or tofu, a serving of brown or white rice, a few lightly steamed vegetables.

Take plenty of vitamin C. Your body needs it to hold the line and stick things back together again as it fights off the infection. When I notice a cold coming on, I take 500 mg a day, and give 100 mg to each of the kids. This usually works, but once a winter, we usually get a cold anyway. When that happens I increase the dosage to 500 mg for

Breakfast
Blintzes for Two*
Hot Spiced Applesauce (see page 96)
Sour Cream
Tea or Coffee

Lunch (Dinner)
Chicken Kiev*
Sweet and Sour Pickles
Brown Rice with Sautéed Mushrooms
Green Salad

Afternoon Snack
Strawberry-Rhubarb Pie*
Tea

Supper
Egg and Lemon Soup*
Rye Bread, Butter
Pears and Cheese

Blintzes for Two
To be served late, on Sunday morning, accompanied by hot spiced applesauce, sour cream, and coffee. Blintzes are actually crepes, cooked on one side, stuffed with cottage cheese, and fried in hot oil. Try the recipe below or, if you prefer, use your own favorite crepe recipe and whatever crepe equipment you may have. I have been making crepes for fifteen years in a small 15 cm (6") cast-iron, well-seasoned frying pan, which I simply preheat; it doesn't require oil. I've also used a 15 cm (6") cast-aluminum omelet pan with sloping sides, but it requires higher heat (sometimes a problem on my wood stove) than cast iron, and, of course, it has to be oiled. My feeling about crepes is that you can make them with practically any pan, provided you have a small vial of vegetable oil handy and a watchful eye.

Rolling a blintz

The Crepes:
Mix up:
    2 eggs
    3/4 cup cold water
    3/4 cup cold milk
    1/3 cup whole-wheat flour
    1/3 cup unbleached white flour or whole-wheat pastry flour
    1 pinch salt
Beat the batter slightly. If you have time, chill the batter for an hour. Don't worry if you don't. Preheat pan and spread around a few drops of light vegetable oil. Pour in 2-3 tablespoons batter and tilt pan so the batter runs around and covers an area approximately 15 cm (6") in diameter. (If you're using a large pan, use a fork to spread it quickly). Cook each crepe until brown underneath and sort of dry-looking on top. Do not turn over the crepe; only one side is cooked. Turn them out on a smooth tray or clean counter, uncooked side down. Leave them like that until you have time to fill each one; they will be reheated later.

The Filling
Mix and mash up:
    1 cup drained cottage cheese or pot cheese
    1 medium or small egg
    1 teaspoon grated lemon rind or 1 teaspoon vanilla
Put one heaping teaspoon of the cottage cheese mixture in a line close to, but not at, one edge of the uncooked side of each crepe. Fold in about 2.5 cm (1") of crepe at both ends of the line of filling, and then roll the crepe tightly. You will end up with a compactly rolled crepe with the ends sealed so the filling will not ooze out. Keep rolling until you run out of crepes and filling.

In a large skillet, heat:
    1/4 cup light vegetable oil.
When the oil begins to crackle, put in the stuffed blintzes, seam side down, but don't let them touch one another. Turn and cook on all sides until evenly browned, about 5 minutes. Drain on brown paper; keep hot in a 250°F (120°C) oven until you are ready to serve, which should be as soon as possible.

Strawberry-Rhubarb Pie
Luscious fruit pies are usually associated with summer, when berries and fruits are fresh and abundant. What could be more warming and special than a tart-sweet pie in the dead of winter. In the late spring and early summer, when rhubarb and strawberries are easy to come by, stew up a couple of quarts of each and jar them or freeze them against that special winter day.

Make the piecrust for apple pie (see page 97). Brush the inside of the bottom crust with a bit of egg white and allow to dry. Meanwhile, preheat your oven to 375°F (190°C)
In a bowl, mix together:
    2 cups strawberries
    2 cups rhubarb
    1 cup sugar
    1 pinch of salt
    4 tablespoons tapioca or cornstarch
Pour into the pie shell and top with second layer, pinching and sealing well. Cut a few steam holes, and place in the oven with a cookie sheet on the rack below—the pie may bubble up a bit before the thickening agent takes hold. After 15 minutes, lower heat to 325°F (165°C) and bake 45 minutes longer. Serve at room temperature.

Egg and Lemon Soup
2 servings

Thick, warm, and lemony, but not sour, this soup is a delicious way to bring something new into your kitchen.
In a heavy saucepan, heat:
    3 cups chicken broth
Meanwhile, break two eggs into a good-sized bowl and beat with a rotary beater or whisk until thick and smooth. Grate into them:
    rind of 1/2 lemon
Squeeze in and beat:
    juice from 1/2 lemon
Then pour in a thin trickle 1/2 cup hot chicken broth, beating all the while with your other hand. Add another 1/2 cup soup, beating to keep the mixture smooth. Then pour the warm egg-lemon mixture into the remaining simmering soup broth, beating and mixing all the while. A wire whisk is the best tool for this, though an egg beater can be used. Take care that the soup doesn't boil. As soon as it is thick, remove from heat and serve at once.

Chicken Kiev
3-4 servings

The end result of this recipe is four to six tender, boneless rolls of pure chicken breast, golden fried and flavored with garlic (or thyme). It's guaranteed to beat any commercially prepared chicken any day.

Preparations for Chicken Kiev should be made well in advance. The rolls should be refrigerated before cooking (this keeps them together in the pan), so if you're going to be pressed for time, do the first part of it earlier in the day.

To begin with, get a fair-sized chicken, 13.5 kg to 22.5 kg (3 to 5 pounds). Remove the legs, thighs, and wings, and use them for other purposes, such as chicken soup (see page 109). Take the skin off the breast, and toss it into the stock pot (see page 110); do the same with the carcass that will be left after you bone it.

Boning Chicken Breast
Starting at the tip of the breastbone, cut down the center, keeping the knife as close as possible to the side of the center bone. Use many strokes, working the meat off gently with your other hand. Then bone the other side, the same way. Each side separates into two filets. Pull up on the smaller, inner pieces, and cut as needed to separate them.

Tenderizing and Subdividing
Lay out each larger piece on a firm wooden surface and lay a piece of plastic or waxed paper over it. Using a rolling pin or the flat side of a heavy knife, pound the meat patiently and gradually out from the center until it's about .6 cm (1/4") thick. Turn it over and pound some more.

With a sharp knife and a steady hand, these filets can be sliced into two thinner pieces. Work the knife close to the side of the meat that had the skin. You don't have to do a perfect job. You also don't have to do the division at all; you will end up with four rolls, two large, two small. Set these aside and lay out each smaller filet, cover with plastic or waxed paper, and very gently pound it out. These pieces are tenderer than the others; be careful, or you'll shred them.

Filling:
The purpose of the filling is to make the rather dry meat juicy and to add flavor to it. You can experiment with the herbs, if you like: try lemon juice and thyme, fresh basil and parsley, tarragon and shallots, or this, the traditional garlic mixture:
Mash up:
   1/2 cup softened butter
Chop fine and add:
   1/4 cup fresh parsley
   2 tablespoons fresh chives
Crush and add:
   2 large cloves garlic
Lay out the filets, and divide up the butter mixture between them, about 1 tablespoon for each small one, up to 2 tablespoons for the larger one. Roll them up and refrigerate, covered, for at least one hour.

To Cook Chicken Kiev
In a small bowl, beat
   1 egg
In another bowl, place
   1 cup whole-wheat flour
Preheat in a deep frying pan
   1/4 cup vegetable oil
Immerse each roll in egg, draining it slightly by holding it over the egg bowl for a moment; then roll in flour, shaking to remove excess flour.

Fry the rolls over moderate heat for about 15 minutes. If you have not split the larger filets, start them first, adding the smaller ones to the pan after 5 minutes or so. Turn as needed as each side becomes golden brown. Serve at once, hot, on beds of rice, accompanied by a good green salad.

Boning a chicken breast

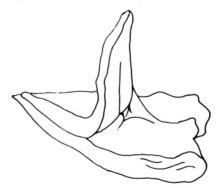

Separating the filets

Cutting the larger filets in half

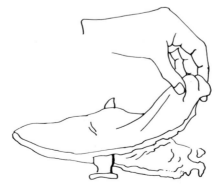

Rolling Chicken Kiev

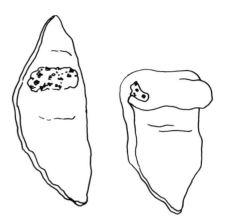

me and 100 mg for them three times a day until we've gotten past it, then go back to once a day for another week. Vitamin C is one vitamin you can't overdose yourself with.

And stay away from cold capsules, narcotics, alcohol. These things just blot out your symptoms and make you feel fine (if a little fuzzy) when you really aren't. You should be listening to your body tell you things, like when you need food, liquid, rest, sleep. Drugs make it hard for you to receive the signals, and in the end make it harder to get better.

Suppose, in spite of your best efforts, you find yourself in the midst of that monumental disaster, the Bad Cold. The infection has now established a beachhead in your head, and is proceeding down toward your lungs, stomach, whatever. You doubtless have a fever, which means your body has to work even harder to fight the thing. You may be hungry, but you can't eat much at one time. You're tired. You hate the world and yourself. Go to bed! Find somebody else to take care of the world. Drink plenty of fruit juice, whenever you want it. Substitute hot, clear soups for meals. When you begin to pick up, go back to light meals, with high amounts of protein and plenty of fresh vegetables and fruit.

Keep Well Stocked

It helps a lot to have foods on hand that are digestible and good for you when you have a cold or other illness in the house. Supplies you can store include:

Fruit Juices and Concentrates: Frozen lemonade, orange juice, and other juices can be kept in the freezer. I like to make and can juices in the summer, using whatever berries or fruits are in season. Simply cook up the fruit in water to cover, until it's soft; then strain through cheesecloth, reheat, sweeten to taste with sugar or honey, and pour, boiling hot, into canning jars. Wipe rim and seal with canning lids. Fruits you can use include raspberries, cherries, cranberries, currants, blackberries, apples; rhubarb, a vegetable, makes a good tart juice as well. These are good cold, but they are especially heartwarming and delicious when served hot; a steaming cup of blackberry or currant-flavored lemonade can soothe you like nothing else on earth. Fruit concentrates and pure lemon juice are also helpful in concocting good hot or cold drinks.

Unflavored Gelatin: Add one packet to ¼ cup cold water and mix; then add to ¾ cup hot fruit juice (see above) and chill. This is just a solid form of juice, though; it should not be thought of as a protein dish.

Wheat Germ, Whole-Wheat Flour, Debittered Brewer's Yeast, Unrefined Vegetable Oil: Use these to make fortified baked foods such as muffins, breads, and light cookies or biscuits. Add a little wheat germ and/or brewer's yeast to hot cereal or any baked goods.

Other foods for fighting colds should be as fresh as possible. (I used to keep a five-dollar bill rolled up in a vitamin C bottle to spend on these things.) They include:

Fresh Fruit: Instant or frozen orange juice is no substitute for the real thing. Buy a bag of oranges and any other fruit you enjoy. Keep it cool and eat as much as you want.

Fresh Vegetables: As you begin to recover, your appetite will pick up and you'll enjoy munching on things like green peppers, carrots, broccoli and cauliflower, Brussels sprouts, cabbage, tomatoes, radishes, celery, celeriac. Eat these raw whenever possible.

Meat: My father's personal cold remedy is a sirloin steak, medium rare, and a big salad. As soon as you're ready for a high-protein meal—or, if you feel you might defeat an oncoming illness by a big jolt of high energy—give it a try. Any lean meat will do as well as steak, chicken, fish, liver, venison, or veal.

Dairy Foods: Eat these in moderation, but don't skip them altogether. An egg in the morning, milk at lunch, and yogurt for a snack will provide vitamins and protein.

Tofu: The favorite tofu for recovery food is Kinugoshi, which has a texture somewhat like yogurt, and a delicate taste that needs no accompaniment.

Whole-Grain Products: If you don't make your own, go on down to the natural-foods store and invest in a loaf of their bread, some low-salt crackers, and other goodies. They may cost more, but remember, you're getting a lot more nourishment—and it's fresh.

Homemade Soup Stock

Soups are wonderful in the winter: they warm the heart, take the chill off your bones, and fill up the hole in your middle. They are invaluable when you or someone you care about is sick. The base of homemade soup, a rich stock, can be made ahead of time and frozen, a particularly handy feature when you are short on energy and/or time. There's no end to the varieties of soups you can make. If you want something thick and hearty, try minestrone, or Scotch broth (page 98) as the basis of a hearty lunch or supper. Serve an exotic, creamy soup like chicken cream (page 111) or egg and lemon soup (page 106) to pamper your sniffles; or start a meal with a good clear soup, like chicken noodle (page 111) or broccoli-tamari (page 113). Homemade soups are better and cheaper than canned versions, and very easy to make.

The Stock Pot: Soups are tastier and better for you when they're based on a good stock. Stock is water in which you've cooked bones, herbs, and vegetables to extract their flavor and nutritional elements. You can pressure-cook stock in a hurry, or simmer it on back of the stove all afternoon, filling the kitchen with lovely smells and making things nice and steamy, good for your sinuses and your houseplants. Later, strain, cool, degrease, and store your stock in the freezer, where it will last for months, or in the refrigerator, where it'll keep three or four days, or longer, if you take it out and boil it every so often.

For example, on Sunday you have Chicken Kiev. You start a chicken stock with the bones and use part of the stock for chicken noodle soup on Tuesday, storing the rest in a tightly lidded jar. Thursday you have lamb chops. Pull out that chicken stock, dump it in a pot, add a couple of liters (quarts) of water and the bones from the chops, and simmer for a Scotch Broth on Saturday. Stocks are nothing if they are not flexible, and you needn't tell all you know; the bones are sterilized by cooking, so you can shamelessly scrape them off plates and into the pot. Another good source of bones is the butcher (including the supermarket one), who will saw up marrow-filled knucklebones for a price, or fill a bag with "dog bones" for free (or almost). If you're on your own, a hatchet is handy for reducing them to pot sizes.

If you have a big freezer and a big stock pot, you can make a few months' supply of soup bases. Stock certainly takes up less room than bones, and it's easy to make a lot. To save containers, line each with a plastic bag and pour in stock. Freeze, then remove the bag of frozen broth; seal and put it back in the freezer. The container is free for other things. Frozen stock will keep in a good freezer bag or container up to six months at $-18°C$ ($0°F$).

Eating Without Meat

I have never liked the word *vegetarian;* people whose diets are meatless rely on many sources of food other than those we commonly refer to as vegetables. Essential among these are grains, nuts, seeds, legumes such as dried beans, vegetable oils, fruits, and, frequently, egg and milk products. There is no essential nutritional substance in meat that cannot be found in plant food. The only real difference, nutritionally, is that fats and protein are more highly concentrated in most meats. This means that if you have been accustomed to eating meat and you decide to go without it, you must pay special attention in planning meals that have enough calories (as well as enough protein) to help you withstand the winter's cold and remain healthy and able.

One of the best plant sources of protein is soybean. It has twice the available protein of most other beans, and four times that of grain. When combined with grains in a single meal, the combination has as much protein as an equal-weight serving of beef. The only problem with soybeans is that they don't taste particularly groovy. In fact, they don't have much taste at all.

Treat soybeans as you would other foods with little flavor or texture interest. For example, substitute cooked, mashed soybeans for pumpkin in pie; even your own grandmother wouldn't know the difference.

Soybeans can be prepared in many ways: as soyburgers, soy grits, soybean stews, and ground up, as a protein additive to baked

---

Home-Made Stocks

Chicken Soup—A Light Stock
Use cooked or, better, uncooked chicken bones. Remove any edible bits of meat, which can be added to the soup later. Leave on or add skin and gristle. Pile into a large pot:
  the chicken bones
  3½ quarts water
  1 whole, peeled onion, stuck with 2 whole cloves
  1 whole scrubbed, unskinned carrot
  1 stalk celery
  1 bay leaf
  1 teaspoon salt
  1 teaspoon vinegar or 1/4 cup white **wine** (to help extract calcium from the bones).
Put a lid on the pot, bring stock to a rapid boil for 10 minutes or so, then lower the heat and cook until the gristle falls off the bones, about 3 hours.

If you're in a hurry, you can pressure cook chicken stock at 6.7 kilograms (15 pounds) for 30 minutes. To cool rapidly, run cold water over the cooker until the pressure goes down. Be careful with pressure cookers full of hot liquid.

After cooking, strain the broth, discarding everything but the liquid. If it has a lot of fat in it, you may want to pour the stock into a tall jar and using a gravy ladle, skim off the yellow fat that rises to the top. Or, chill it; the congealed fat is easy to remove from the top of the jelled broth. Most people enjoy a little fat in their soup, though, so don't be too thorough. If refrigerated in a tightly lidded container, stock will keep up to five days. It may be used in sauces or casseroles as well as chicken soups or other soups calling for a mild flavored base, such as egg and lemon soup (p. 106).

Note: Although chicken necks are traditional ingredients in chicken stock, I would like to warn you against using them if they come from commercially grown chickens. The antibiotics and steroids chickens are shot up with these days tend to accumulate in the necks, making them the repository of a lot of unpleasant additives we can all do without.

The stock can also be made, as described, with turkey, duck, goose, or rabbit bones.

Beef Stock—A Strong Stock
Use cooked or, better, uncooked bones from beef. Remove any edible bits of meat to add later to your soup. Put bones in a pot, along with:
  3½ quarts of water
  1 whole peeled onion
  2 whole scrubbed unpeeled carrots
  1 bay leaf
  1 pinch thyme
  1 tablespoon salt

  1/4 cup vinegar or 1 cup red or white wine
Cover the pot, bring to a boil, then lower heat and skim any scum that has risen to the top. Simmer 4 hours. (The vinegar may make an odd smell as it reacts with the calcium. Don't worry about it; the soup will taste and smell fine when it's done.)

If you're in a hurry, use a pressure cooker—6.7 kgs (15 lbs) pressure for 1 hour ought to do it. Be careful to lower pressure **completely** before removing lid by running cold water over the pan.

Next, strain the broth, discarding everything but the liquid. If it's very fatty, pour into a tall container and skim off the fat with a gravy ladle, or refrigerate until the fat hardens on top and remove it. You need not get out every speck; a little fat is good in a soup.

Use meat stock wherever strong soup base is called for, as in minestrone (p. 96) or Scotch Broth (p. 98). You can also use meat stock in sauces, gravies, stews, casseroles, or as the cooking liquid for grains such as rice or bulgar, substituting stock for part or all of the water called for. Remember that your stock has some salt in it, so decrease salt slightly when you are using it in a recipe; or adjust seasoning to taste at the end.

Note: The same procedure using lamb, pork, or venison bones will give you lamb, pork, or vension stock.

Menus for Getting on Top of a Cold

To fight off a cold, or when you're on the road to recovery, you should follow a diet such as you would to lose weight: high in protein, vitamins, and minerals, low in animal fats and rich foods. The only difference is you don't reduce your caloric intake.

Breakfast
Fresh Grapefruit or Orange
Boiled Eggs on Whole-Wheat Toast
Milk or Cheese

Tea or Snack
Hot Apple Cider
Laban with Rye Crackers*

Lunch
Broiled Calves' Liver
Brown or White Rice
Raw Alfalfa Sprouts with Yogurt Dressing*

Dinner
Chicken Cream Soup*
or Beef Noodle Soup
Whole-Wheat Crackers
Pear and Cheese

If you're quite sick, you probably won't feel like eating at all. It's better to have a series of snacks, planned to meet all your nutritional needs without adding anything unnecessary.

Breakfast
Hot Whole-Wheat Cereal with Wheat Germ
Milk

Elevenses
Hot Lemonade with Fruit Concentrate
Cottage Cheese
Chopped Lettuce with unrefined Safflower oil and lemon-juice dressing

Lunch
Broccoli-Tamari Soup (see page 113), or Quick Chicken Noodle Soup*
Whole-Grain Crackers

Tea or Snack
Fruit Compote
Arrowroot Biscuits
Hot Herbal Tea

Supper
Egg Custard or Yogurt with Honey
Hot Cider

Evening Snack
Hot Grape Lemonade*
Arrowroot Biscuits or Bran Muffin

Yogurt Dressing for Salads
For those of you who like a creamy salad dressing, but find mayonnaise cloying or boring after all these years, try using yogurt as your base. Here are some additions to try alone or in combinations:
　garlic
　celery seed
　curry powder
　dried mustard
　tarragon
　lemon juice
　tamari soy sauce
　sesame oil
　ginger (grated or powdered)
　scallions or shallots
　cayenne pepper
　minced parsley, coriander, or mint

Laban
Even if you hate yogurt, you will love this stuff. It's also a painless way to get some fresh garlic into your system—garlic is a super source of vitamin C, something you need lots of when you're feeling down.

Take a square of clean rag made from an old sheet; a 25 cm - 30 cm (10" - 12") square is about ideal in size. Dump a half-liter (pint) of yogurt in the middle and gather up the four corners. Fasten with a strong piece of string so the yogurt is enclosed in the sheet. Hang in an out of the way place—a cupboard pull provides a good hanger—with a bowl underneath to catch the drips, and leave for at least 12 hours.

The stuff that drips into the bowl is whey, and it's worth keeping if you want to try to make a sour rye bread without using sourdough starter. Open the sack and scrape into a bowl the now quite cheesy drained yogurt. Chop fine 1 clove of garlic, or put it through a press.

In the unlikely event that you can find fresh mint in the winter, chop up 3 tablespoons. If you can't, use 1 tablespoon of dried mint.

Combine the garlic and mint with the drained yogurt and let stand for at least 1 hour—the longer it stands, the more the flavors develop and blend.

Spread on crackers as though it were one of those expensive creamy spiced cheeses.

Chicken Cream Soup
4 servings

In a saucepan, heat
　2 tablespoons butter
Sauté until limp:
　2 shallots or 2 tablespoons minced onion
　2 stalks celery, chopped
Add and heat, covered:
　4 cups chicken stock
In a measuring cup, mix:
　4 tablespoons cornstarch
　1/2 cup chicken stock
Add this to the hot broth and turn down heat; simmer 10 minutes.
Add:
　1 cup minced cooked chicken
　optional: 1/2 cup cream
Simmer 5 minutes longer, being careful not to allow it to boil. Just before serving garnish with 3 tablespoons chopped fresh parsley.

Quick Chicken Noodle Soup
4 servings

In a saucepan, heat
　2 tablespoons vegetable oil
Sauté until limp:
　1 stalk celery, chopped
　1 carrot, chopped
Add and bring to a boil
　4 cups chicken stock
Then add:
　1/2 cup egg noodles
　1/2 cup peas or chopped spinach (can be frozen)
　optional: 1/2 cup chopped cooked chicken
Bring to boil again and turn down the heat until it just bubbles. Cook, covered, about 10 minutes, until noodles and vegetables are tender. Season to taste and serve at once.

Hot Grape Lemonade
2 cups

In a small teapot, mix:
　3 tablespoons honey
　2-3 tablespoons freshly squeezed lemon juice
　4 tablespoons Welch's grape juice or home-made grape juice (or any other flavor you may have cooked up)
　1½ cups hot water
Stir until honey dissolves.
Can also be served chilled.

goods like cake, muffins, cookies, and breads. And, if I can let you in on a not-so-well-kept secret, so-called meat extenders sold are little more than flaked soybeans. There's no reason why you can't mix up your own meat-extending concoction next time hamburgers, meatballs, or meatloaf is on the menu. Season well.

Chinese and Japanese cooks have been experimenting with soybeans for several thousand years, and have come up with a number of different products that retain the high protein and vitamin component of soybeans. Among these are soy sauce and miso (a soybean paste), which are made by fermenting the beans. These store well and are widely available in natural-food stores, markets that carry Oriental foods, and, increasingly, the neighborhood supermarket. Another product, tofu (bean curd), is made by soaking, grinding, and pressing the beans to make a soy "milk," out of which tofu is separated, like cheese curds, and lightly pressed. Fresh tofu has a pleasant, mild flavor and delicate texture, but is quite perishable. It should be stored wet in the refrigerator and used within three days of purchase. There are many varieties of it available in Oriental food markets and some natural-food stores, particularly in large cities. Try it fried and served with sauces or stir-fried with vegetables, for winter eating; it's also good in soups, sandwiches, and salads.

For those who, like myself, do not have ready access to fresh tofu, there is still hope. It can be frozen and then thawed when you are ready to use it. This changes the texture somewhat, but it is not a change for the worse. The tofu becomes porous, considered an advantage in China (where thawed tofu is a winter treat) because the flavors of accompanying sauces are absorbed more quickly. If you want to try frozen tofu, buy some fresh whenever you get the chance and are within a day's travel from your freezer.

You can also make your own tofu. It's a slightly complicated procedure and has a reputation for being fraught with failure, but *The Book of Tofu,* by William Shurtleff and Aikiko Aoyagi, will show you the way if you are interested.

Finally, there's some magical stuff called dried bean curd sticks, which lasts practically forever. It is, in fact, tofu thoroughly dried and brittle. To use, soak in hot water until the sticks soften—ten to fifteen minutes—then cut into bite-size pieces. Reconstituted dried bean curd is not exactly like fresh tofu, by any means; rather it is a thing in itself and very delicious. Pick up a package next time you are near a source of Chinese provisions, and try it either stir-fried, in soups or braised dishes, or cold with sesame oil and a crunchy vegetable to contrast textures.

Another good thing to know about is the combinations of grains, beans, and nuts or seeds that, when eaten together, provide a good deal more usable protein than they would if eaten separately. Such information, and much more, can be found in Frances Moore Lappe's classic *Diet for a Small Planet.* If you are interested in meat-less eating, don't make a move without reading this book.

In the colder months, you will do well to use a lot of oil in cooking and baking. Remember that oils have twice as many calories per gram as protein or carbohydrates, and that the oil will stick around in your system, providing warmth and energy for many hours longer than a dish of rice or a bowl of soup. Remember, too, that some foods are particularly high in oils, such as nuts, seeds, nut butters, eggs, and whole-milk cheeses. In cold weather, it's particularly important to include oils in your breakfast and lunch foods, so you

Meatless Menus

Breakfast
Oatmeal
Roasted Sunflower Seeds
Dried Fruit
Tea

Snack
Sesame Seed Cookies*
Hot Cider

Lunch
Broccoli-Tamari Soup*
French Baked Beans*
Carrot and Sprout Salad with lemon juice
and safflower oil dressing

Dinner
Deep-Fried Tofu with Sauce*
Mixed Stir-Fried Vegetables*
Brown Rice

Broccoli-Tamari Soup
6 servings

Tamari soy sauce is a wonderful soup flavoring, delicate and tasty. Vegetables like carrots, onions, beets, beans, Brussels sprouts, and cauliflower are great in it, but my favorite is broccoli.

In a heavy 2-liter (2-quart) pot, heat:
    2 teaspoons vegetable oil
Sauté in the oil:
    1 small onion, sliced thin
Add and bring to a boil:
    4 cups water or broth
Reduce heat to simmering; add:
    2 cups broccoli flowerets
    1 cup peeled broccoli stems, cut in 1.3 cm (1/2") slices
Simmer until broccoli is crisp yet tender, 5 to 10 minutes. Just before serving, add:
    3/4 cup Tamari soy sauce
    1/2 teaspoon unrefined sesame oil
Do not boil after adding these ingredients. Serve at once.

French Baked Beans
6 servings

A slightly different way to make beans, with a delicious garlic flavor. Serve with roasted sunflower seeds to complement flavor and increase protein.

Soak overnight in 4 cups water:
    2 cups white or navy beans
At least 3 hours before serving, heat in a heavy 3-liter (3-quart) pot:
    4 tablespoons corn oil
Chop and sauté:
    1 medium onion
    1 carrot
    2 stalks celery
Mince and add:
    2 cloves garlic
Sauté until onions are limp, stirring to make sure garlic doesn't burn. Drain the beans, discarding the soaking water. Add to the pot, along with:
    2 cups water
    1 teaspoon salt
    1 bay leaf
    1/2 teaspoon freshly ground pepper
Bring to a boil, then lower heat and simmer, covered, for 2 hours.
Add:
    3 potatoes, diced
    1/2 teaspoon basil
    1/2 teaspoon thyme
    1/2 cup tomato juice or 1 cup fresh or canned tomatoes
    1/2 cup celery, chopped
Simmer until potatoes are done.

Sesame Seed Cookies
36 cookies

Preheat oven to 350°F (175°C).
In a heavy frying pan, roast 1 cup sunflower seeds 5 minutes over moderate heat, stirring as you brown.
Add and continue to stir for another 5 minutes 1/2 cup sesame seeds.
Take care not to let them burn
Allow seeds to cool. In a large bowl, beat:
    1 egg
Add and beat in:
    3 tablespoons corn oil
    1/4 cup honey
    1/2 teaspoon vanilla
    2 tablespoons water
Add and mix well:
    1/4 cup whole-wheat flour
    1 cup rolled oats
    1/4 cup currants
    1/4 cup raisins
    Roasted seeds
Spread in a thin layer on an oiled 30 cm (12") square cookie sheet, and press flat with wet hands. Bake 10 to 15 minutes, or until browned. Cut into 36 5cm (2") squares with a wet knife and cool before removing.

Deep-Fried Tofu

There are a number of ways to deep-fry tofu; this version is crisp and delicate, but it must be served at once. Allow two cakes of tofu per person.

To prepare tofu for frying, leave it out of water but in the refrigerator overnight. An hour before frying, cut the tofu into 1.3 cm (1/2") thick slabs and lay them on a folded dish towel; cover closely with another towel. When you are ready to fry, dip each slice in beater egg, let the excess drip off, and then dip in arrowroot, cornstarch, or kuzu. Gently shake off excess powder, and proceed with next slice.

Use vegetable oil (any kind) and, if you have it, add a few ounces of sesame oil for extra flavor. (You can keep the oil to reuse several times if you strain it and store it in a lidded jar after each use.) Heat the oil in a wok, heavy skillet, or electric frying pan, to a temperature of 350°F (175°C). If you don't have a candy or frying thermometer, you can gauge the temperature with bread cubes. When a cube sinks only slightly and rises at once, the oil is hot enough. Keep your oil clean as you fry; between batches, skim out stray scraps and allow the temperature to rise again before putting in more tofu.

Fry about six slices at a time until golden brown; turn once. Drain on paper and serve at once. Pass around a dipping sauce made of:
    4 tablespoons Tamari soy sauce
    4 tablespoons sherry, sake, or 2 teaspoons honey and 4 tablespoons water
    2 teaspoons lemon juice
    2 tablespoons grated radish or horseradish
This should be enough sauce for four people; increase proportions to serve more. If you prefer a thickened sauce, heat the ingredients gently and pour in a mixture of:
    1/2 cup water
    4 teaspoons arrowroot, cornstarch, or kuzu
Simmer and stir until thick, about 5 minutes.

Mixed Stir-Fried Vegetables
4 servings

The main thing about this sort of cooking is that it should be done very quickly; as the Chinese say, you should "surprise" the vegetables. Prepare your ingredients before heating up the pan, and use fresh, crispy vegetables for the most part. Some frozen vegetables, such as corn or peas, can be added during the steaming stage. The end result should be a tasty mixture, each vegetable cooked just long enough to tenderize it. Serve with rice, bulgar, or millet, and 1 cup sunflower seeds, roasted as described in the recipe for Sesame Seed Cookies. A serving of grain and stir-fried vegetables (including bean sprouts) has about 10 grams of protein, roughly half the amount you need in a full meal. The deep-fried tofu adds the rest, as well as providing a pleasing taste and texture contrast with the crunchy fresh vegetables.

You can use any of a variety of vegetables. Here is my favorite combination:
    3/4 cup onion, chopped medium fine
    2 cups shredded cabbage
    1½ cups thinly sliced carrots (cut on a diagonal to get oval-shaped rounds)
    1/2 cup thinly sliced winter squash
    1 cup mung bean sprouts
    1/2 cup soybean sprouts
In a heavy wok or frying pan, heat 3 tablespoons mild vegetable oil until oil crackles when a drop of water lands in it. Dump in carrots and squash; cook 5 minutes, stirring occasionally. Push these out to the sides of the pan and add onions and cabbage. Cook and stir 3 minutes. Then add soybean sprouts and:
    3 tablespoons water
    2 tablespoons Tamari soy sauce
    1 tablespoon white or brown sugar
Cover pan at once and steam 3-5 minutes, until carrots are almost tender. Add mung bean sprouts and steam 1 minute longer. Serve immediately.

will remain healthy and enjoy your days outdoors as much as you would at any other time of year.

Brown Bagging

Once I saw a brown bag made for an emperor's son. It consisted of two lacquered boxes and three small lacquered bowls with closely fitting covers, a neat little jug with a lid you could drink out of, and two chopsticks inlaid with mother-of-pearl. Everything fitted together and was strapped into a silken case, embroidered with butterflies and dragons. It was probably tucked into his sleeve as he ran out to play in the royal gardens, and was doubtless filled with tasty tidbits of things that are just as good eaten cold as any hot luncheon or supper in a boring old dining room.

Preparing food to eat away from home is not a new culinary challenge. There are endless possibilities. The contemporary favorite, sandwiches, can be made with dozens of combinations of different foods. Unfortunately, many are tired old combinations. For a change, try cold chicken, home-baked muffins, a grain salad (cold rice or bulgar, chopped vegetables, and a light oil and lemon juice dressing). For dessert, there are cookies, fruit, yogurt, dried fruits, nuts, egg or rice custard, and fruit-nut loaves. (You don't eat all this at once, of course; I'm just trying to point out that you don't have to live your life in a perpetual round of peanut butter or bologna sandwiches.)

Of course, when you're away from home, there are always restaurants, lunch counters, sandwich bars, and so forth. Whereas some of the restaurant foods offered to the public should be avoided at all costs—particularly deep-fried things, like chicken, fish, chips or fries, and doughnuts, since the odds are overwhelming that the fat in which those things are fried is rancid, and it just isn't worth the health and flavor risk—there are many foods you can buy that warm the heart and provide a welcome interlude on days when you are away from your own kitchen. If you're eating out, choose soups, lean meats, salads, cheese, and other dairy products. There's no reason not to bring your brown bag along too, and order something hot and steamy to enjoy along with your homemade goodies. Come to think of it, maybe that would wake up restaurant proprietors a bit; why can't *they* provide good things like homemade whole-grain bread, sprouts, yogurt, and a variety of cheeses?

Brown Bag Ingredients

If you're making sandwiches, start with a good, fresh bread, one that holds up well in a sandwich and has a firm, chewy texture. Whole-grain breads are better for you than those made with refined flours; not only do they have more B vitamins, but they also provide roughage to help you digest other things in your meal. Rye bread is a favorite for many sandwiches because of its firm texture. Whole-wheat bread is good too, but it should be made with eggs and milk, or you may find it too crumbly for a sandwich that has to do some traveling.

If baking your own bread is out of the question, buy bread fresh from a bakery or natural-foods store. Avoid chemicals and preservatives added to "prolong shelf life." They won't prolong *your* life, and they make boring sandwiches.

Sometimes it's hard to find decent things to drink when you're away from home. Avoid using this as an excuse to grab an easy can of

Brown Bag Specials

Whole-Wheat Sprout Bread
2 loaves

In a large ceramic bowl mix:
    2 cups warm water
    2 teaspoon dried yeast
    1/4 cup nonfat milk
    2 teaspoons salt
    1/4 cup vegetable oil
    1/4 cup wheat germ
    1/4 cup molasses
    1 egg
Allow to sit in a warm place for 5 to 10 minutes; then take it down, beat it a bit, and add:
    3 cups whole-wheat flour
Beat with a wooden spoon until your arm gets tired; cover with a damp cloth and set to rise in a warm place (70°F, 21°C) for an hour or so.

When you have a free minute, spread some wheat flour on a board and prepare to knead the bread. But first, add to the batter:
    1½ cups wheat sprouts (see page 104)
Then beat in, mix in, and finally knead in:
    3-4 cups flour
Keep adding flour until the well-kneaded dough is smooth and even-textured. Set to rise until doubled in bulk (about 1½-2 hours), punch down, and shape into loaves. Raise in greased loaf pans until doubled; bake at 350°F (175°C) for 45 minutes. Remove from pan as soon as it is baked; cool before slicing.

Sandwich Ingredients

What makes a tasty combination of sandwich foods tends to be a personal matter. I used to go out with a guy who was crazy about peanut butter and sardines on Westphalian pumpernickel. Here are some more likely combinations; many of them include sprouts, which to my mind hold up better than lettuce; they don't get limp and slimy and have a welcome, nutty crunch. (Instructions for growing sprouts can be found on p. 104)

Dijon Special

Spread a thin layer of Dijon mustard on rye bread. Top with Swiss cheese or sliced cold meat, or both. Cover with half an inch of alfalfa sprouts, and another slice of rye. Some like mayonnaise spread on the second slice, some don't.

Fresh Cheese Salad

Spread a slice of whole-wheat sprout bread (above) with cream cheese or creamed cottage cheese mixed with 2 tablespoons chopped green olives. Top with a healthy layer of mung bean sprouts, then the second slice of bread.

Nuts and Honey

There are so many nut and seed butters to choose from: sesame and sunflower seeds, almond, cashew and the old favorite, peanut butter. Instead of jelly, top with a thin layer of honey, or, to save teeth, molasses. Good with whole-wheat bread, sublime if you add a layer of thinly sliced bananas.

Bean Sandwich

Sure you can. Spread a thick layer of mashed leftover beans on Anadama bread. Top with a little chili sauce, a layer of grated cheddar cheese, a bunch of alfalfa sprouts; close up with second slice of bread.

Curried Egg Salad

A lively variation on an old theme. Chop up two hard-boiled eggs. Bind together with mayonnaise flavored with a teaspoon of curry powder, about ten raisins (better if you plump them in hot water for a few minutes and then drain), and a tablespoon of toasted sunflower seeds. Spread on whole-wheat bread, top with alfalfa sprouts, and cover with second slice of bread. Try it with chicken salad instead of eggs next time you have leftover chicken. Yum!

Molasses-Oatmeal Cookies
30 cookies

Rich and chewy—and they hold their shape in a brown bag.

Preheat oven to 350°F (175°C).
Mix together in a large bowl:
    1/2 cup vegetable oil
    1 beaten egg
    1/4 cup molasses
    1/2 cup dried milk
    1/2 cup brown sugar
Then add, all together:
    1/2 cup wheat germ
    1 cup whole-wheat flour
    2 cups rolled oats
    1 teaspoon baking powder
You can do it all in one bowl if you pile dry ingredients on top of wet, then mix up the dry ones lightly with a fork before stirring from the bottom. When they're all blended together, wet your hands and roll balls with about 1½ tablespoons of batter for each one. Flatten between your palms and bake on a lightly oiled cookie sheet 10 minutes.

Peanut Butter Cookies
30 cookies

Rich, delicious, and addictive, these can be made with either smooth or crunchy peanut butter.

Preheat oven to 375°F (190° C).
Mash together in a shallow bowl:
    1/2 cup brown sugar
    1/4 cup vegetable oil
    1 cup peanut butter
Mix in:
    1 egg
    1/2 teaspoon salt
    1/2 teaspoon vanilla
    1½ cups whole-wheat flour (or slightly more, as needed)
Roll the dough into small balls, about 1½ tablespoons each. Place on oiled cookie sheet and flatten with a fork, or press flat with your thumb. Top each mound with an unroasted peanut.
    Bake 15 minutes, but keep a very close eye on them after 10 minutes. When they are lightly browned on top, take them out and let cool 10 minutes before attempting to remove them from the cookie sheet.

soda pop. Vacuum bottles are the obvious solution; you can bring whatever you like, hot or cold, from home. Another solution is to pack a lot of wet and crunchy accompaniments to your sandwich to make up for the lack of drink. Fresh raw vegetables in plastic wrap or a plastic bag, a piece of fruit—both are eminently portable. Mix up some seasoned salt—salt, dried parsley and/or dill, a dash of cayenne pepper, some celery seed—and take it along to season your vegetables. Or season-salt a carton of yogurt, close it back up, and use it as a vegetable dip.

When it comes to fruit, apples and bananas are delicious, but you don't have to stick to those old favorites. How about some pineapple—fresh, canned, or dried? Melon, berries, peaches, and other wettish fruits can travel if you have a plastic, closable tub or plastic bag. Pack some toothpicks to spear bite-size pieces and you're in business.

Cookies? Of course. But make them yourself, out of healthful ingredients. Everyone will thank you for that.

When you pack a brown bag, think of what you'd like to eat (or feed to your loved ones) and then figure out a way to get it on the road. Be resourceful. You'll be amazed at what qualifies as road food when lots of napkins go along for the ride.

At this writing, one of the heaviest storms we have had in several years is battering the windows, and drifting so deeply that we wore snowshoes this morning just to get to the barn. The storm was even worse for most of the northeastern United States, and there's another monster brewing out in Saskatchewan. I listened to an excited man last night on the radio telling how he served hot drinks and sandwiches to over two hundred people in his house last night. He was on his way out to get more coffee.

In some ways, winter is a lot of fun, if you're ready for it. Part of being ready is understanding how to cope with the stresses of cold, wind, and snow. But there is another factor, a sort of willingness to change your path, to meet each situation freshly with whatever you have at hand. This applies equally to dealing with the energy problems of your house, your car, and your body. In the winter, you need more calories; but you also need ingenuity.

An old Cape Breton story exemplifies both that needed ingenuity and a knowledge—perhaps instinctive—of the staying power of fat- and protein-rich foods. A man was about to set out on a sleigh ride into town. It was a long, cold ride—24 kilometers (15 miles)—and as he was leaving, his wife took some hot baked potatoes out of the oven. She instructed her daughter to cut open the potatoes and slip a thick wedge of cheese into each one, wrap them up in clean cloth, and tuck them into the robes that covered his feet. The daughter went along with the hot potatoes, but she couldn't understand what the cheese was for.

"Well," said the woman, "when he gets to town, he'll eat the potatoes; that'll be his lunch."

"And what will keep his feet warm on the way home?"

"Ah," said the wise old woman. "That's where the cheese comes in."

Some Special Sources

There are enough cookbooks available in stores, libraries, and on the shelves of friends to keep us all reading (and maybe cooking) through many a winter. I won't attempt to provide even a partial list. But here are a few books that I have found helpful, and that relate to the things we've been talking about in this chapter.

Brown, Edward Espe. *The Tassajara Bread Book*. Berkeley: Shambhala, 1970. If you are new to bread baking, this can be your most valuable tool. Brown explains what happens when you make bread, and somehow that makes it all possible.

Lappe, Frances Moore. *Diet for a Small Planet*. New York: Ballantine, 1971. This is *the* book if you're thinking about eating without meat, or reducing your meat intake, for economic, philosophical, or ecological reasons. There are some fine recipes for meatless eating, but the main attraction is the theory and practice of complementing proteins.

Restino, Susan. *Mrs. Restino's Country Kitchen*. New York: Quick Fox, 1976. My cookbook for all seasons, not just winter. There's lots more about bread, stocks, soups, plus information on making cheese, preserving fruits and vegetables, smoking meats, and plenty of recipes for simple and fancy dishes.

Shurtleff, William and Akiko Aoyagi. *The Book of Tofu*. Berkeley: Autumn Press, 1975. Everything you wanted to know about tofu, including how to make it, with lost of recipes for using it. Yes, it is a rich enough subject to fill a whole book.

Growing Things

Winter is surely not a hospitable season for flowers and plants. Outdoors, only the evergreens exhibit signs of life; and indoors, plants and people alike fight against the ill effects of overheated and bone-dry rooms in the dreary winter light. But it need not be so. Growing things can work wonders on the spirits, and they require in turn only a bit of special winter care.

Plants in the Indoor Winter Environment

Green, growing things give a room the feeling of cozy comfort. Flowers, especially when out of season, are heart-warming, and when there is an inner glow the cold outer world seems somehow less forbidding.

Several options are available to us for brightening our rooms with plants and flowers. Always, everywhere, there are cut flowers. And there are flowering plants. I think a heavily budded potted plant just coming into bloom is one of the best bargains on the market today. For about the same price as a pair of steaks, you can bring home a growing, blooming plant—an azalea, a kalanchoe, a cyclamen— that will last for three or four weeks if cared for properly. Of shorter duration but spectacular, and somehow more homey, would be a pot of forced tulips or primroses in flower, or a cineraria with its gaudy display of gypsy colors.

Perhaps you want something more permanent. You should consider florists' amaryllises; African violets and other gesneriads (my favorite is the hybrid Streptocarpus, with its colorful flowers that keep on coming week after week); clivia; free-flowering cacti such as Thanksgiving, Christmas and Easter cactus, and especially the newer crab cactus hybrid; and the numerous other blooming plant species that can be counted on year after year.

Then there are the tropicals. Most of these are non-blooming but have showy foliage. Winter is the time the ficuses, dieffenbachias, philodendrons, and the like go to my plant room (where light is very bright, humidity very high, and temperature moderate) and also the day the summer slipcovers come off the furniture and winter rugs go back down. I relegate the ferns as well to the plant room or green-house because, to me, a fern says cool, woodsy dell, which is the last thing I wish to think of in the dead of winter. In the winter I want flowering things everywhere: clusters of pots of forced bulbs in front of the French windows; African violets, minature begonias, and fairy

One-time Flowering House Plants

These flowering plants are "disposables." Buy top-quality specimens just coming into flower (direct from the production greenhouse, if possible, to save them from exposure to the frequently hostile conditions of the florist shop) and maintain them as directed. When flowers fade, toss them out. (Save clay pots, of course.) It is tempting to keep a faded cyclamen, or kalanchoe or gloxinia—surely it will bloom again!—but they rarely come back to full beauty without greenhouse conditions. You'll get more than your money's worth the first time around; buy new plants to replace faded ones.

Begonia x cheimantha (commonly, but incorrectly, called *B. socotrana* and hybrids)—Christmas begonia: Several fine cultivars in white, shades of pink, and rosy red, blooming from top to bottom. Average room temperature; humidity as high as possible; bright light but no sun; soil evenly moist.

Calceolaria herbeohybrida—pocketbook plant: Jungle colors, mostly reds and yellows; jungle form, pouched and strange. Cool to cold room temperature; humidity as high as possible; strong light, no sun; soil evenly moist, never soggy.

Campanula fragilis—blue bellflower: Small, trailing plant with splendid sky blue flowers top to bottom. Hang in a cool to cold east or north window; humidity as high as possible; soil evenly moist.

Campanula isophylla alba—white bellflower; Treat as *C. fragilis* above.

Crossandra infundibuliformis—crossandra: Lovely four-inch spikes of irregular, salmon orange flowers above handsome, shining dark green leaves. Cool room temperature; humidity as high as possible; bright light with some sun; water generously, then allow top of soil to nearly dry before watering again.

Cyclamen persicum—florists' cyclamen: Waxy, butterflylike flowers hover over spreading foliage that is often beautifully marbled. Cool to cold room temperature; humidity as high as possible; strong light but no direct sunlight; soil evenly moist.

Euphorbia pulcherrima—poinsettia: The Christmas flower. In shades of red, greenish white, and rather dirty pink; flowers are small red and yellow affairs at twig ends, with brilliant, leaflike bracts surrounding them to make the "blossom." Average room temperature; humidity as high as possible; bright, diffuse light, no sun; soil evenly moist.

Hydrangea macrophylla—Hortensia hydrangea: In hues of pink, blue, or white, with huge soap-bubble heads of sterile florets delicately tinted; a true Easter flower. Cool room temperature; humidity as high as possible; bright, indirect light, no sun; soil evenly moist (these take a lot of water). Buy when blossoms are greenish and allow to develop color in a chilly north window.

Kalanchoe blossfeldiana—kalanchoe: Strong stems bearing heads of vivid brick, salmon, even mustard or lavender flowers, leaves dark green, red edged, and rubbery. A fine plant that lasts for months. Average room temperature; brightest possible light, but sun only through a sheer curtain; drench with water and rewater only when soil surface is dry to the touch.

Lilium longiflorum—Easter lily: Several strains of this superb lily come from florist shops; buy plants with only one open bloom and lots of buds. Cool room temperature; strong light (an hour or two of morning sun is good); water copiously and allow to dry somewhat (soil surface not powdery, but barely moist) before watering again.

Pelargonium x domesticum—Lady Washington (or pansy) geraniums: Scores of named varieties are offered by specialists for late winter and early spring. These bloom only once. Cool to cold room temperature; bright light (some sun); drench, then allow soil to dry on top before watering again.

Pelargonium x hortorum—zonal-leaf geranium: These are the common geraniums, in many colors and forms, all beautiful. Cool to average room temperature; very bright light, some sun; water copiously, then allow soil to dry on surface before rewatering.

Primroses

Primula malacoides—fairy primrose or baby primrose: Flowers in tiers, leaves often silvery green, a charming small plant with sweet fragrance and flowers mostly in shades of pink. It was once very common but has been virtually driven from the market (as have other primroses) by the hot, dry air of modern houses. Grow at cool to cold room temperature; humidity high as possible; bright, sunless light; soil evenly moist.

Primula obconica—top primrose: From a rosette of roundish green leaves rise clean stalks with showy heads of round flowers; white to dark red, but mostly pinks and lavender. Culture as *P. malacoides*. Note: Leaf hairs sometimes cause dermatitis when handled.

Primula sinensis—Chinese primrose (and var. *stellata*—star primrose): Small, starry flowers, the latter deeply cut, often frosted foliage; a gem of a tiny plant. Culture as above.

Saintpaulia ionantha hybrids—African violets: Grows in many forms and colors, but basically purplish blue to pink. A central cluster of violetlike flowers surrounded by uncommonly handsome, more or less spoon-shaped leaves. If people would enjoy these while beautiful and then discard them rather than worrying about how to keep them going, they would love them more. Cool to average room temperature; humidity as high as possible; strong, diffuse light (no sun); soil constantly moist, but allow to dry somewhat at ten-day intervals.

Senecio cruentus—florist's cineraria: The gayest, brightest of all florist's plants. Large "maple" leaves spreading below a huge, flat cluster of brilliant daisy-type flowers, usually banded white, with purple, lilac, cerise, rose, pink and many other shades. New sorts may be monotones of brick, salmon, apricot, and similar delightful hues. Strictly a one-time-around plant, so buy when just coming into flower. Cool or chilly room temperature; humidity as high as possible; bright light with a bit of sun; soil always moist. (These take a lot of water where humidity is less than ideal.)

Sinningia speciosa—florist's gloxinia: Large, oval, velvety leaves lie flat, surrounding a central cluster of huge, bell-shaped flowers, in white, imperial purple, intense ruby red, or speckled, mostly edged white. Average room temperature; **very high** humidity; strong, filtered light (sunny window with a sheer mesh curtain); soil constantly damp.

Solanum pseudo-capsicum—Jerusalem cherry: Grown for the large, shining orange fruit scattered over the leafy green bush, these go on and on when properly positioned. Cool room temperature; humidity as high as possible; strong, filtered light; soil constantly moist.

Streptocarpus x hybridus—cape primrose: Long, strap-shaped, lettuce green leaves, with ruffled and wavy bell-shaped flowers in white, pink, rose, blue, or purple, usually with throat stripes. Although fragile in appearance, they are actually strong plants that bloom for a long time. Cool room temperature, humidity as high as possible; average, filtered light; soil moist, but allow to dry somewhat every ten days to two weeks; give dilute house-plant fertilizer biweekly for continuous bloom.

primroses on end tables; amaryllises on the window sills or next to the table lamps at night when windows are frosty. But that's a personal preference; I have friends who love every green, growing thing. In winter they fill the library, the entry hall, the living room, even the dining room and kitchen, with monumental specimens of tropical foliage plants, and then spread around on tables, windowsills, and in stray corners my sort of clutter of flowering plants. The result is marvelous: when you walk in it's like a midwinter trip to Bermuda.

What You Can Do for Your Plants

Plants, unlike most animals, are immobile; they occur in nature in an environment that is conducive to their proper development. Accordingly, most plants are not terribly adaptable. Each has more or less rigid requirements for light intensity and duration, temperature (often cooler at night than in the day), humidity, soil moisture, soil fertility, and many more factors. The very first thing you can do for your plants, winter or summer, is to become familiar with the environmental specifications of each and then meet these as closely as possible.

Look over the various habitats you have to offer. Quite likely your bathroom is warmer, steamier, and possibly, brighter than any other room in the house. A seldom-used bedroom may be bright and cool, and free of artificial lighting most of the time; that is, it offers a long-night, short-day situation during the winter months. Your living room may be as arid as the Sahara Desert unless you are an antique furniture buff or a musician with a grand piano, in which case the humidity surely runs 40 percent or more. Once you have a list of the various habitats in your home, you are in a position to select plants that will thrive on what you have to offer. Let's take a look at the major environmental factors that affect plant growth.

Humidity

The absence of water vapor in the air of modern heated dwellings is the greatest bugbear to indoor plant culture. Even in the desert, dew forms at night because the air cools and there is sufficient humidity near the ground to condense. We almost never find dew on

---

Landscape with Your House Plants

Group your house plants into eye-pleasing compositions. If you intend the groupings to be relatively permanent, be sure the plants you chose are environmentally compatible; that is, that they share temperature, humidity, and light requirements.

For example, in a relatively bright entry hall with a blank wall, start with something tall and columnar—a ficus, an aralia, brassaia (florists' schefflera), or other plant with bold foliage. Complement it with a broader, finer-leaved species, preferably with foliage of quite a different hue—a dizygotheca, araucaria, croton, or dracaena, perhaps. Complete the base of the grouping with an assortment of species that contrast greatly in leaf size, form, texture, and color, such as aspidistra, jade plant, grape ivy, English

ivy, kalanchoe, peperomia, or wandering Jew. Perhaps such an out-and-out tropical look is not your taste, and you would prefer something that looks regional and summery. If you can offer strong light; high humidity; temperatures of 17° - 18° C (62° - 64° F) day, 13° - 15° C (55° - 58° F) night; and no drafts, a tree-form citrus underplanted with a grouping of shrubby pittosporum, rhododendron, and flowering maple and surrounded by potted ferns and pots of forced spring bulbs, cyclamen, begonias, and the like, will give a very different feeling to the hall.

Even on a table, working on a handsome (and waterproof) tray that ties the composition together, you can achieve a little picture with, say, a small Natal plum, a dwarf pomegranate, and a clay saucer planted with nertera. A different picture would be primroses, African violets, and English ivy, perhaps in one of its color-leaved forms.

Move your plants about. Even if they are environmentally incompatible, they'll be all right for a few hours. Make a composition to greet your guests; surprise them with a display of plants in the bathroom; or decorate a sideboard or a mantlepiece with a luxurious array of green and flowering plants.

Build your own light table

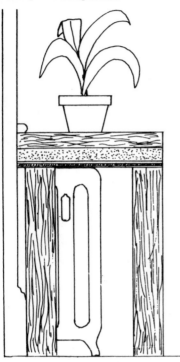

Don't let plants hang over edge of table

Table should be as low as possible

Plywood can just clear radiator

Table should extend 6'' or more beyond radiator

Back of table should be tight against wall or windowsill

If the best light for your plants happens to come from a window over a radiator, don't despair. By building an insulated radiator cover, you can keep the radiator on to heat the room and still be sure not to boil any tender roots. The trick is to sandwich a layer of styrofoam between two layers of wood. The bottom layer gets hot all the way through, but the styrofoam absorbs virtually all the heat. The table top absorbs the rest.

Construction is as simple as a table can be, four legs and a top. The best material to use is ¾'' plywood. Anything thinner is risky—potting soil soaked with water can get heavy, especially when you discover that more and more of your plants like to take the winter sun. As illustrated here, the table top has a raised edge made of 1'' x 3'' molding, but it is a not entirely necessary refinement. If your table top and underside pieces are to be identical they can be cut with the same saw cuts. If you make a raised molding, remember to make allowances for its thickness all the way around the table top, if you want the sides to be flush. You can hide the styrofoam by running a molding around the table that covers everything from underside to table top. Sink your nails through the molding into the sides of the table. It is important to keep the bottom insulated from the top. Even with no molding around your sandwich, the parts of the table top will need nothing but friction and their own weight to keep them in line.

Stretch your table as far as the room available, even beyond the end of the window. Some of your plants will enjoy the indirect light and relative warmth. If your table gets longer than about 4½ feet, you should add a center leg to keep it from sagging. The simplest way to attach the legs is to screw them on from the top side of the bottom layer, but there are prefabricated legs that come with their own systems for attachment. If the legs seem wobbly, brace them with diagonals—it's better to be safe than to risk spilling your favorite plants. Stretch the available space by building out beyond the radiator as far as you can afford to, but remember that no matter how far out you go, the heat will still be strong enough to bake your stems if they hang over the edge.

A table top with a molding can be made still more effective if you put water in it. The first step is to make it watertight. This can be done by caulking the inside carefully and then painting it with many coats of a waterproof paint—deck paint now comes in a variety of interesting colors. Be sure that all cracks are completely sealed. Strew decorative pebbles or marble chips to the top of the molding, then add your water. Sit your plants just above water level—never in the water, or their roots will rot—you won't need saucers. The humid atmosphere created by the evaporating water will make your plans even more comfortable.

—Alan Ravage

Table can be as wide as you want, even wider than window.

Add middle leg if table is longer than 4½ feet.

Leg can be braced, if necessary, for lateral strength.

our tables and chairs, even when we lower the night temperature by ten degrees or so in our living quarters. No doubt that is a good thing—from the furniture's standpoint; but it indicates that the air is pretty dry for plants.

In a room heated by hot air or by steam or hot-water radiators, the relative humidity can frequently be as low as 10 or 15 percent. But there are ways to raise the humidity throughout a room to make it a more congenial place for plants, and it is relatively simple to increase humidity locally. If your house is heated by a hot-air furnace, you can install an automatic, and adjustable, humidifier in the cold-air return. Ceramic, plastic, or metal devices of various designs can be fitted into or behind radiators and filled with water, which freely evaporates as the radiators heat. Look for these in hardware and houseware stores. In the kitchen, you can let a tea kettle simmer daylong; a kettle on a hob in any hot fireplace will produce a steady flow of steam and strain your gas bill not at all. If you have radiators directly below windows, you can have metal trays fitted over the tops of the heat units; fill them with largish pebbles, add water, and set plants in clay saucers directly on the pebbles. Water vapor from the warm water below will bathe the foliage even though the humidity in the rest of the room will remain rather low.

One thing that does *not* work is misting plants daily. Not that a fog of fresh moisture is harmful; it's simply that such a bath is too brief to meet a plant's humidity requirements.

Healthy Air

Visit almost any commercial greenhouse during midday, and you will notice that even when the temperature is far below freezing the ventilators will be slightly open to freshen the air. All plants thrive on fresh, gently flowing air and, conversely, they deteriorate in stuffy, motionless air. That's one side of the story. The other is that strongly moving air is not suitable for plants. They cannot take drafts of hot air from a register nor chill breezes from a nearby door that opens and closes from time to time. What is best for plants, in fact, is just what is best for you: cool room temperature and fresh air. You can provide this through a window, opened slightly at the top, if possible, and shielded behind a sheer curtain, or from the air gently swirled through the room (some of it pulled in from outside via minute cracks around windows and doors) by a burning fireplace. Be aware, though, that air brought in from outside in winter, when warmed to room temperature, is invariably dry as zweibach; tend to your humidity!

Water Around the Roots

All green and growing plants take water from the soil for various cellular activities, especially photosynthesis. But plants grow less in winter because days are shorter and light is dimmer, temperatures are lower, and air is drier. During the winter months all plants use less water than during the rest of the year; some actually go dormant or nearly dormant. Begin a gradual cutback in your watering after the autumnal equinox. Learn to click a coin—always the same coin—against the rims of your sound (uncracked) terra cotta pots to recognize the moisture level. A clear ring means pretty dry, a very dull thud indicates soggy; there is a nice gradient of tones between these extremes. By late January, the moisture level in the soil should be reduced to a bare minimum. Then start a gradual—very gradual—increase to reach full growth watering sometime in April or May.

A great deal has been written about bringing water to room temperature to avoid jolting plants severely with a drench of cold water. As a matter of fact, highly technical research indicates that for quite a few species the temperature of water used on pot-grown plants makes little or no difference. Who hasn't heard of conveniently watering hanging baskets by dumping in a few ice cubes and leaving them to melt gradually? At the farm, where our water comes pure and unadulterated out of a 600-foot (180-meter) hole in the ground at about 14° C (58° F), I water my plants right out of the hose winter and summer. At the house in town, I draw water for the plants into a pair of two-gallon jars and let it rest for a couple of days. It isn't temperature I'm concerned about, it's pollution. The chemical smells of the city water disappear, and at certain times of the year sediment of something indescribable (I am afraid to ask experts what it is) collects on the bottom. After its two days of "breathing" the city water seems relatively free of chlorine and stuff, and I pour it on the plants.

When possible, always water on a rising temperature. Morning applications invariably are better than nighttime ones because a plant uses far more water during light hours (for photosynthesis) than in darkness; it hardly seems fair to subject your plants' root systems to a cold, soggy night for the sake of affording yourself a few extra minutes in the sack in the morning. Roll out and give those plants a drink—but only if they need it—of properly aerated water.

Fertilize Correctly, If at All

House plants stay green and some few bloom in winter, but it is the rare species that makes any appreciable amount of leafy growth. Plants respond to the shorter day, cooler temperature and drier air by going into a drowsy state. Most need no fertilizer at all in winter. If fertilizer is applied the plant may respond by breaking out of its quasidormancy only to perish, or if no growth occurs the additional soluble chemicals in the soil may injure the roots, throwing the plant into a decline. If you are in doubt about whether or not particular species should be treated with fertilizer, follow this rule of thumb: Don't fertilize from late summer through early spring. And don't think that spring has sprung until the plant has sent up a new shoot or two, indicating that growing time has arrived and that a bit more fertilizer and water would be appreciated.

There are a few plants that thrive on small amounts of certain fertilizers in winter. All are blooming sorts. Any of the gesneriads, begonias, primroses, citrus species and cultivars, hibiscuses, cyclamens, and such—note that all are more or less cool-to-cold-weather plants—can do with very diluted fertilizer solution at watering time, but not every time you water. For the plants in the house, I mostly use a chemical formulation; Plant Miracle, Ra-Pid-Gro, Shultz Instant, and several of the Peters formulations are on my shelf. Or I use tea-colored manure water, made by steeping a dry but not weathered cow chip in a gallon or two of water for a few days. When relatively fresh, the solution is not overly pungent. In the greenhouse and in the basement (where I keep plants under flourescent light) I use any of the above, or, more frequently, half-strength fish emulsion solution (diluted with twice as much water as is recommended on the bottle). It stinks, but the plants love it.

Please note that I have not mentioned "plant food" anywhere; this is nothing more than careful English. Plants make their own food —basic carbohydrates—by combining water and carbon dioxide in the presence of light. The process is called photosynthesis. Fertilizers supply chemicals needed for various metabolic processes; they make growth possible, but they are not food.

Control Insect Pests and Diseases

Indoor plants look seedy when infested by various pests or speckled with disease. There was a time when newly purchased tropicals, raised locally under glass by skilled horticulturists, came to us clean and healthy; today many plants grow in subtropical fields, accumulating an assortment of exotic ills. By and large they are not serious outdoors, but when the affected plants come into our houses, where the environment is less than ideal, the pests and diseases take over. I have seen a beautiful, and apparently healthy, supermarket hibiscus go down under a massive encrustation of scale in just three months, and I have had flower stalks of Florida-grown amaryllis topple and fail due to leaf-scorch disease, which came with the bulb. Many of us move our house plants to the garden for the summer to reduce daily maintenance and to give them a new lease on life. When they come indoors in the fall, spider mites, various scale insects, mealy bugs, and a host of other creatures, as well as some disease organisms, ride along. In the balmy atmosphere, removed from natural controls, these pests and diseases might fizzle out or they might explode into a destructive plague.

When you bring home a new plant, unless it comes from a florist you know sells only clean stock, isolate it from your other plants for

---

"Permanent" Flowering House Plants

Some potted, winter-flowering plants can be kept going year after year with a little or a lot of gardening skill and effort. Among the most reliable and easiest to care for are:

Abutilon x hybridum—flowering maple (and *Abutilon megapotamicum* var. *variegatum*).
Acalypha hispida—chenille plant
Ardisia crispa—coral berry

Begonia
B. rex, B. masoniana, B. albo-picta, and others are grown for their showy foliage.
B. boweri, B. fuchsioides, B. manicata, and others are grown for their winter flowers.
Beloperone guttata—shrimp plant
Brunsfelsia calycina—Brazil raintree

Cactus
Schlumbergera bridgesii—Christmas cactus; *S. gaertnerii*—Easter cactus; *Echinopsis multiplex*—Easter-lily cactus; *Zygocactus truncatus*—Thanksgiving cactus.
Easy-to-grow and colorful hybrids of some of these have been recently introduced. Pot up rather crowded in equal parts garden loam, old cow manure or compost, sand, and damp peat, adding ¼ cupful bonemeal per peck of soil.

Grow damp through summer months, but allow the soil surface to dry before watering again in winter. Apply dilute house plant fertilizer solution monthly, from early summer until plant is in flower. From fall until in flower grow in bright light, in a cool room, **with no light at all after normal sundown.** Most are short-day, long-night plants, which will not set buds if exposed to artificial illumination.
Carissa grandiflora—Natal plum
Clerodendrum thomsonae—glory-bower or bleeding-heart vine.
Clivia miniata—scarlet kafir lily
Gibasis geniculata—bridal veil tradescantia
Hibiscus rosa-sinensis—Chinese hibiscus

Hoya
H. bella; H. carnosa—wax plant and its many forms.
H. purureo-fusca and others grow as thick-leaved, twining creepers. Average room temperature; high humidity; very bright light (with sun); garden loam with a bit of coarse compost or old strawy manure, sand, and coarse charcoal. Keeping soil on the dry side in summer, and somewhat drier in winter, usually yields good bloom.
Impatiens walleriana—patience plant, sultana, busy lizzy, and other species and many hybrids.
Medinilla magnifica—medinilla
Nematanthus (Hypocyrta) wettsteinii—goldfish plant, and other species.

Oxalis hirta, O. lasiandra, O. rubra, and others.

Pelargonium
P. crispum—lemon-scented geranium; *P. x domesticum; P. graveolens*—rose-scented geranium; *P. x hortorum; P. peltatum*—ivy geranium; *P. tomentosum*—peppermint-scented geranium and many others.
Puncia granatum—pomegranate, and var. *P. nana.*
Rhododendron obtusum—florist's Kurume azalea (many sorts).

a month or two and examine it frequently for indications of disease and pests. Take similar precautions with plants brought indoors after a summer in the garden. I move my plants indoors in stages to gradually acclimatize them to new conditions. The first step is to the outer edge of the porch; then to the inner area, against the house, where there is less light, stiller air, and dew at night; and finally, inside. When they reach the porch, I scrub the pots, give plants a thorough grooming so dead and damaged foliage and stalks are removed, prune some, and spray all at least twice, some three times, at five-day intervals to rid them of pests. *Never* bring a plant into the house from the garden as is, because eventually you will have a dreadful mess to clean up.

Examine your house plants frequently all winter. I don't pretend to know how mealy bugs get on an African violet that has been indoors and healthy for years, but they do. It is a mystery to me how scale insects find their way into the house to attack permanent residents such as an English ivy, a ficus, or an aralia, but they do. Diseases seldom become a major indoor problem if one follows a proper management program of limited water, no fertilizer, strong light, and low—for the particular species—temperature.

Plant Grooming

Make it a habit to pick up your plants and handle them every

Keeping House Plants Healthy

Proper cultural methods that produce healthy plants—soil moisture and aeration, fertilization, humidity, light, and ambient temperature management—are basic. Beyond these, two aspects of plant health are important.

Sanitation is your first line of defense. It is essential that you use only clean containers free of snails, slugs, millipedes, and other plant-chewing creatures that hide in soil. Prompt removal of damaged, diseased, or weak leaves or shoots is part of sanitation, as is removal of spent flower stalks. In short, sanitation means keeping the entire plant and its container clean. This is largely hand work.

Disease and insect control comprise the second aspect. Disease is usually not a serious problem inside the house. Various leaf spots sometimes develop. Treat by quickly removing infected leaves and spraying the plant with a mixture used to control black-spot-mildew-rust on roses. If mildew appears (it looks rather like flour dusted over leaves), wash plants with fresh water, spray as above, improve air circulation, and increase the strength of light. Sometimes leaf tips or margins appear burned. This is most frequently environmentally stimulated by overly dry air, overwatering, or too much soluble chemical (fertilizer or something from the water) in the soil.

Although theoretically any garden insect can be a problem indoors, the serious troublemakers are:

Aphids: Tiny, soft-bodied, green, pink, gray, or blackish, wingless or winged, nearly immobile insects visible on new shoots or emerging buds. Aphids suck plant juices and weaken or destroy the growth on which they are found. Very vulnerable, they can be controlled by spraying with (a) aerosol insecticide from a house plant bomb; (b) a

relatively "safe" insecticide, such as diazinon or malathion; (c) washing repeatedly with a strong jet of rather warm water. Repeat treatment at least three times at five- to seven-day intervals. Option (c) is rarely completely effective.

Mealy bugs: Adults appear as white, cottony masses, with nothing recognizable as an insect showing, in tight places such as leaf axils, bud sheaths, leaf undersides, and so on. Immature mealy bugs are tiny, creeping, nearly transparent greenish, flattened insects that move mostly on undersides of leaves and stems. Mealy bugs are very difficult to eradicate, so get at it and keep at it. With a cotton-tipped swab dipped in rubbing alcohol, wipe away all visible cottony masses; then spray, as described above for aphid control: the same chemicals, the same frequency, but four to six applications are usually necessary.

Scale insects: These come in a variety of shapes and colors. On house plants they appear most commonly as immobile masses resembling tiny oyster or clam shells or fish scales. These are adult forms. Using a hand magnifying lens, look along veins on leaf undersides and you will see nearly transparent, flat, immature forms migrating to spread the infestation. Scale control means all-winter chemical warfare in most cases. Use chemicals as described for aphid control, adding a teaspoonful of Ivory Flakes per gallon of solution. Spray thoroughly to run off all plant surfaces and stems. Repeat at five- to seven-day intervals for at least six applications.

Spider mites (red spider): These are not insects, but arachnids. Look for nearly microscopic creatures and bits of webbing on leaf undersides; leaves commonly develop a mottled, yellowish appearance due to killed cells. Procedures recommended for control of aphids, mealy bugs, and scale insects should be tried for spider mite infestations. Adding soap to the spray solution is beneficial. Change your chemical at each applica-

tion (shift back and forth between two or rotate with several), as mites quickly develop a tolerance to a particular chemical.

Note:
An easy way to manage smallish plants is to prepare your chemical solution in an enamel or plastic container (a plastic wastebasket is suitable) reserved for that purpose only. Put on rubber gloves, pick up your plant, invert it, and swish it around in the solution, dipping right to the soil line. Set it aside to drain and go to the next plant. Dispose of excess solution by pouring it away on a gravel driveway, around the foundation of a building, or on a porous terrace surface (brick in sand, for example). If you're a city dweller without a yard, pour it into the soil around a sidewalk-planted tree. Do not pour down a sewer or storm drain, or where rain could puddle it so that birds or pets could be affected.

few days. Pick off dead and discolored leaves, spent flowers, and dry, scaly bits of tissue from stems. From time to time use an old table fork to loosen the soil, which gets packed down from water. With a plastic scouring pad remove green scum from the outsides of pots and the white and brown mineral encrustations that build up around the pot rim and soil line. If leaves look dull and lusterless, give them a gentle rubbing with a piece of old nylon stocking; I am anti-leaf wax because I think it clogs the stomata (air-exchange pores in leaves), but I don't like to look at water-spotted and dusty leaves either. Sufficient natural wax is usually present to give a healthy glow to glossy leaves, if you supply the elbow grease. Practice good horticulture by knowing and keeping track of your plants' Latin and cultivar names. That plant isn't just any old ivy geranium; it is *Pelargonium peltatum* "Comtesse Degrey." The difference between a rank amateur and a pro is knowledge!

What Your Plants Can Do for You

Ornamental plants are beautiful things. Beauty of form; graceful foliage, often handsomely marked or colored; and delightful flowers that may also be sweetly scented are among the attributes that make a plant ornamental. Of course, beauty is in the eye of the beholder. I have a dear friend who is a genius with cacti. She grows a

Showy House Plants Grown for Foliage

Almost all foliage house plants are tender, woody species (a few are fleshy succulents) that tolerate average room conditions. Hundreds of species are available; unfortunately, many of the handsomest ones are conservatory or greenhouse plants that soon deteriorate in the house. The following list includes some of the most foolproof sorts, but there are many more.

Agave americana—century plant, and many other species.

Aloe vera—medicine plant, and many other species

Aloysia triphylla (Lippia citriodora—lemon verbena.

Araucaria excelsa—Norfolk Island pine.

Asparagus meyeri—Meyer's asparagus; *A. myriocladus, A. plumosus*—fern asparagus; *A. sprengeri*—Sprenger's asparagus.

Aspidistra elatior—cast-iron plant, and var. *vareigata*.

Brassaia actinophylla (Schefflera actinophylla of commerce)—Queensland umbrella tree.

Bromeliads—*Aechmea, Billbergia, Canistrum, Cryptanthus, Dyckia, Guzmania, Hechtia, Nidularium, Puya, Tillandsia, Vriesia,* and several more genera.

Chamaerops humilis—Mediterranean palm.

Cissus rhombifolia—grape ivy; and other species.

Citrus—*Citrus aurantiifolia*—lime; *C. aurantium*—Seville orange; *C. limonia*—lemon; *C. paradisi*—grapefruit; *C. sinensis*—sweet orange. There are several other species, varieties, and hybrid forms. The ornamental Calamondin orange is *C. mitis;* kumquat is *Fortunella japonica.*

Codiaeum aucubaefolium; C. spirale; C. variegatum var. *pictum* (with many cultivars)—all known as croton.

Cordyline australis—Dracaena indivisa of florists; *C. terminalis*—red dracaena, many cultivars, some the ti plants of commerce.

Crassula argentia—jade plant, and several other species.

Ficus—Ficus benjamina—weeping fig; *F. deltoidea (F. diversifolia)*—mistletoe fig; *F. elastica*—India rubber plant and var. *decora* among others; *F. lyrata*—fiddleleaf fig; *F. microcarpa*—laurel fig (here belong the nitida and *F. retusa* of commerce); *F. pumila (F. repens)*—creeping fig, with several varieties; and several other fine, ornamental species more or less scrambled in commerce.

Hedera helix—English ivy, with hundreds of ornamental forms.

Laurus nobilis—laurel, bayleaf (of the kitchen).

Lithops bella—stoneface—one of the many "living stones"; also applies to related genera. These are tiny succulents.

Malpighia coccigera—miniature holly.

Peperomia caperata—emerald ripple peperomia; *P. obtusifolia*—pepper face; and many more species, varieties, and cultivars.

Pereskia aculeata—lemon vine, a leafy climbing cactus.

Philodendron—species of philodendrons, some 200, vary from herbaceous clumps to fleshy, fibrous climbers and treelike forms. Most are rain forest plants, ill-suited to average house conditions. Some handsome "self-heading" or bushlike hybrids are fairly adaptable to indoor culture, as are *P. hastatum, P. oxycardium* (florists' *P. cordatum*), *P. selloum,* and *P. wendlandii.*

Pilea cadierei—aluminum plant; *P. involucrata*—panamiga; *P. microphylla;* and others.

Pittosporum tobira—Japanese pittosporum.

Plectranthus australis—southern spurflower (often misidentified as Swedish ivy); *P. oertendahlii*—Swedish ivy; and others.

Pleomele thalioides—pleomele (often sold as dracaena); and several other species.

Podocarpus macrophylla—Buddhist pine, and its varieties.

Polyscias balfouriana—aralia; *P. filicifolia*—

fern leaf aralia; *P. fruticosa*—variegated aralia; *P. guilfoylei* var. *quinquefolia*—celery-leaved panax.

Sansevieria thysiflora—bowstring hemp; *S. trifasciata*—snake sansevieria, mother-in-law's tongue; var. *laurentii,* with yellow lateral stripes on leaves; and var. *hahnii,* a dwarf "bird's nest" form.

Schefflera, see *Brassaia actinophylla.*

Senecio macroglossus var. *variegatus*—variegated wax vine; *S. mikanioides*—parlor ivy, Germany ivy.

Tolmiea menziesii—piggyback plant.

Tradescantia fluminensis—wandering Jew; see also *Zebrina pendula.*

Zamia floridana—coontie, and other cycads in this and other genera.

Zebrina pendula—wandering Jew. Several other genera carry this common name; many are first-rate ornamentals.

couple of thousand of them and thinks each is lovely, a special thing. I can agree that a silken, colorful cactus flower is among the most beautiful blossoms in creation, but I wouldn't dream of living with those spiny, awkward-shaped plants the year 'round. She, on the other hand, has never accepted my offer for rooted cuttings of my fragrant geraniums. Regardless of your taste in plants, without a doubt you will find some that delight you. They become friends to greet you when you enter the room.

Plants are a wonderful form of recreation; with the barest minimum of information on what to do and, more important, what not to do, you can grow a beautiful specimen. Soon you may be into collecting a group of plants, into propagation, perhaps even into plant breeding. It is rather nice to know that a living, growing thing depends on you for its well-being but is sufficiently undemanding to let you go off for the weekend. Any time one gets seriously interested in something—stamp collecting, macramé, guppies, flowers on the windowsill—it is a healthy thing because it takes one's mind off oneself (and futile frustrations due to the latest iniquities of politicians). I do not subscribe to all the current ballyhoo, expressed in glowing generalities and charmingly innocent of supporting scientific data, which claims miraculous mental cures as the result of taking up plants. It is called horticultural therapy. I suspect that getting involved with a tank of baby angelfish would be just as beneficial. But I do agree that messing about with plants every day is a pleasant and profitable pastime, more so because the plants tolerate a remarkable amount of abuse and misunderstanding, and still grow and blossom.

Many plants smell good. Elizabeth tilted her aristocratic nose at hyacinths (in her *German Garden*), but welcomed their heavy scent indoors during the long Pomeranian winters. The Midwestern climate precludes fine outdoor displays of hyacinths in my garden, but I make up the difference indoors. They are one feature of my timed sequence of scents. For the fall and early winter months I like voluptuous, rich "inside" scents: rose geranium in its several forms, lemon geranium, peppermint geranium, and others of this pungent group sit handily on windowsills and in the entryway, offering their leaves for rubbing. Lemon verbena, indoors for its brief visit before going to the greenhouse (it gets terribly tatty-looking inside), suggests lemon time on the Amalfi Drive. The citrus trees suited for indoor cultivation are even more bitter-pungent, but the leaves have to be pulled and crushed to release the scent, which does nothing for their looks. Pineapple sage, *Salvia rutilans,* is another of my favorites. It comes indoors in November bearing long spikes of slender, madder red, sweetly fragrant flowers, and the pinched leaves smell just like pineapple. All of these scents and more are foreign to this part of the world; they suggest nothing of the garden, but of indoor gardening. In addition, several of my orchid plants hark back to the early breeding program for fall-blooming hybrids at the Missouri Botanic Garden, and some of them are delightfully scented.

Pine cones, evergreen boughs of all sorts, holly, and English ivy take the spotlight beginning with the Advent wreath and reaching a crescendo with the Christmas tree and all the rest of it on December 24. These are plant materials, not plants, it is true; nevertheless, the fresh greenery brings the fragrance of an evergreen forest to the house. It all goes on January 6, but by then we are ready to think of scents of spring. Winter has become a bore, even though there are weeks of it to come. It's a good time to begin the succession of forced hyacinths; polyanthus narcissus "Grand Soleil d'Or," "Chinese Sacred Lily,"

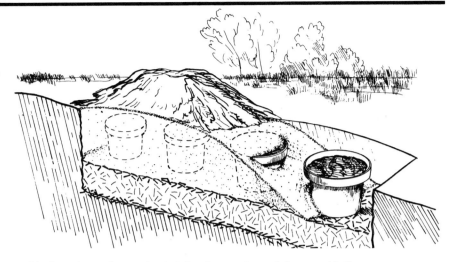

Tulips, hyacinths, squills, chionodoxa, grape hyacinth, pushkinia, crocuses, and most other hardy (frost-tolerant) spring-blooming bulbs, corms, and tubers can be forced into early flower. The technique is specific; unwarranted shortcuts invariably lead to failure. Pay attention to each step and to temperature.

Buy best quality, display-bedding-grade bulbs, corms, and tubers that are sound, free of blemishes or injury, and absolutely dormant. If you buy these early, hold in chilly storage (the refrigerator will do fine) until at least mid-October. Pot up all before November 15.

The containers for use may vary; the best are the squattest terra cotta flower pots known as bulb pans. A drain hole is essential; deep containers hold moisture and roots decay.

For rooting medium you may use bulb fiber (see To Force Once-blooming Tender Bulbs, page 130) or a soil mixture of equal parts garden loam, coarse sand, and coarse brown peat, with a generous amount of charcoal chips added (**not** from barbeque briquettes, which contain toxic binders) to sweeten the soil. Most commercial soilless potting media—Promix, Redi-Earth, and others—are suitable.

Potting a bulb

Procedure: Cover the hole at the bottom of the pot with a curved piece of broken clay pot and lay on several more pieces to insure fast drainage. Toss in crumbly damp potting mix to bring the potting medium halfway up the container. Insert fingers a few times to *lightly* compact the medium. Adjust level of medium so the nose (tip) of the bulb will be slightly below the rim of the pot; corms and tubers should be completely covered to once their depth. Set in bulbs, corms, or tubers (one variety per container), with ½.. (1.3 cm) or so between each and between the sides of the container. Add more growing medium, sifting it carefully around each bulb so air is excluded—air pockets are ruinous. Cover bulbs so only noses are exposed; cover corms and tubers completely. Growing medium should end about 1/2" (1.3 cm) below the rim. Finish

with clean, dry sand up to rim. Label each pot.

Set planted pots in a tray of water that reaches halfway up their sides and leave them to soak for half an hour to an hour. Remove from the water pan and drain, but water immediately once or twice from the top to insure uniform saturation. Set pots in a cool 7°- 13°C or 45°- 55°F dark place for seven to ten days, during which time roots will form.

In the garden dig a rather deep trench; toss in 6" (15 cm) or more gravel, chat, or coarse sand to act as drainage. Set the pots on this, pouring clean, damp sand around and over each. Arrange pots from one end of trench to the other in the order in which they will be removed for forcing. When all pots are covered with 1" (2.5 cm) of damp sand, add soil to a depth of 6" (15 cm), water well, add the remainder of the soil and mound the excess in a broad crown to shed rainwater. The pots need to be deep enough so bulbs will not freeze; in very wintery places, make the trench in a coldframe that can be closed later, or wait until surface soil is lightly crusted with frost (to insure a chill below) and mound it with a great pile of brush or compost.

After pots have cooled and bulbs, corms, and tubers have become vernalized (at least eight weeks; ten to twelve weeks is better), dig out a few pots. Scrub them clean and pour out the excess sand. Water once, and hold in the dark at 6°- 9° C (42°- 48° F) until shoots appear. Move to strong light (no direct sunlight), still cool. When foliage is well developed, increase temperature slightly. Temperature through forcing and blooming should never exceed 17°- 18° C (62°- 65° F) for best quality, texture, and duration of bloom. Cooler is better. A good place to display forced spring bulbs, corms, and tubers is in a "flower bed" put together in a waterproof tray filled with damp peat (protect the floor or carpet below with paper and plastic) surrounded with old bricks. Such a display in front of a French window or glass door takes advantage of the cool, bright position. Bring in more pots every two weeks.

Tuberous rooted ranunculus and anemone hybrids perform well with this regimen, yielding unbelievably luxuriant foliage and wonderful, colored flowers.

To force hyacinths in water, see **To Force Once-blooming Tender Bulbs.**

A trench for potted bulbs

Forcing paperwhites

Hippeastrum (Amaryllis)

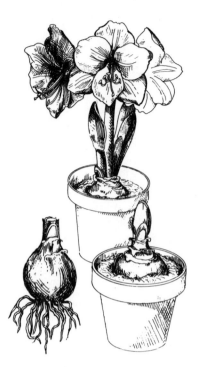

To Force Once-blooming Tender Blubs

To Force Reblooming Tender Bulbs

Paper white narcissus and their relatives, "Chinese Sacred Lily" and "Grand Soleil d'Or" are tender (not frost-tolerant) daffodils easily forced and discarded. Buy hard (not spongy), unbruised bulbs in the fall and store in the refrigerator vegetable bin until six to eight weeks before you need flowers. "Plant" in shallow bowls of pebbles, water to the bulge of the bulb at first, but only to the base when roots are well developed. Alternately, you can plant them half-buried in commercial bulb fiber, or make your own (see below). Keep in a dark, cool, (7°-13° C (45°- 55° F) cupboard until rooted and sprouted. Move to strong north or east light in a chilly, unused room until flowers open. Warm temperature will yield limp leaves and papery, poor quality flowers.
To make bulb fiber mix:

   6 parts loose, coarse brown peat
   2 parts crushed oyster shell (from feed store)
   1 part crushed (not powdered) hardwood charcoal

To use, dampen, toss to fluff, dampen again, and then allow free water to drain away. Keep bulb fiber moist, not soggy. It produces better foliage and flowers than pebbles do. Discard to compost heap when bulbs finish blooming.

Amaryllises (hippeastrums), cyrtanthus, sprekelia, vallota, and several orther sorts of tender but durable bulbs add greatly to the indoor garden during the early months of the new year. Most of these are easily kept potted the whole year, and need repotting only when they are bursting their bounds, since all bloom best when tightly potted and somewhat root-bound.

Buy only the best quality bulbs from specialists; avoid supermarket bargains and magazine ad come-ons; inferior stock invariably performs badly.

Pot up new bulbs September through October. Equal parts garden loam; dampened, coarse brown peat; clean, coarse sand; and well-decayed cow manure or leafmold; with ¼ cupful steamed bonemeal and 2 cups hardwood charcoal chips (**not** barbecue briquette chips) per pot make an ideal potting mixture. Select a pot that is 1" - 2" (2.5 cm - 5 cm) larger in diameter than the bulb. Fill it one-third full of broken crock; add a thin layer of peat screenings or sphagnum moss fibers and then the potting soil, **lightly** compacted. Plant hippeastrum shallowly, just to the bulge of the bulb; cover most others to about two-thirds their height, but plant cyrtanthus more deeply, with just the necks exposed. Water well, and store in a cool (9°- 13° C, 48°- 52° F), very dark place (a closet, fruit cellar, or in a box in the garage), where the bulbs will root. Watch for signs of top growth. When it appears, move to stronger light and water sparingly, increasing water as the bud or foliage, or both, develops. When buds show color, or when foliage expands, fertilize with half-strength house plant fertilizer solution every three to four weeks. Cut away spent flower stalks at the base and maintain leafy plants in vigorous growth with ample water, occasional fertilizer, and strong light. The

leaves will be making nutrients to be stored in the bulb for next year's blooms. When weather warms, move plants to a bright (no direct sunlight), windless part of the garden and continue monthly fertilizing and frequent watering until the onset off fall weather.

As nights begin to cool, discontinue fertilizing and begin to reduce watering. Before the first frost, bring plants indoors and lay pots over on their sides in a dark place. Withhold water and fertilizer. When foliage yellows, cut it away at the base and store the pots in a cool, dark, dry place to rest for at least eight weeks; ten weeks is better.

When bulbs become extremely crowded, do not repot until it is time to wake them up for winter blooming (after the dormant period). Proceed as above, potting each bulb separately. Small offset blubs may be potted several to the pot to grow on to blooming size.

and "Paper white"; fairy primroses (and others); pots of scented English violets, which come in briefly then go back to the deep cold-frame, since they cannot tolerate house heat; cyrtanthus; and other spring-scented species. Of course, there are many unscented plants in the background, too, but I wouldn't give a hang for an indoor garden that failed to perfume the air.

Some of the best forcing bulbs are not especially fragrant. Tulips, daffodils, and hybrid amaryllis (these really are hippeastrums) have only a little scent, some have none. So it is with Spanish squills, crocuses, grape hyacinths, chionodoxas, and other commonly forced spring bulbs and corms. But they are so very beautiful.

Forced dormant branches bring spring indoors early and really are a heart-warming sight. Again, this isn't indoor gardening in the strictest sense, but it will tend to brighten the dullest days of late winter. The technique is simple provided you supply the proper environments. You need a place to hold the container of branches while buds swell and break. It should be bright and airy, with a day-time temperature of 17°-18° C (62°-65° F), and a night temperature of 10°-13° C (50°-55° F), with humidity as high as possible, especially during daylight hours. A spare bedroom may serve for this; I find a windowed corner of the basement laundry room ideal. An inexpensive room humidifier can be a big help during the day. With the place ready and waiting, fill a deep bucket with tepid water, take your sharp clippers, and head for the garden. It is best if the temperature is above freezing, but I have brought in frozen branches and thawed them under a cold shower.

The Healthy Aspect of Anticipation

What can be more stimulating, more exciting (and warming) than the tickling anticipation of a delightful event? The idea that spring is coming can by itself warm us through winter's fading weeks. But there is much more to excite us. We order seed catalogs in November and can barely wait for them to begin to arrive. When they have, we dig out last year's orders and garden performance records as a basis for a current "want" list and add to it species we have heard of since the last order. Then we sit down, evening after evening, dividing up the order among the companies we like best, rather like Napoleon hovering over and dividing up a map of Europe. With the orders in the mail it's an antsy time until seeds begin to arrive, but we fill in those days cleaning up seed flats and pans, rounding up germinating mixtures and ingredients for making our own, and laying out a reasonable schedule for planting seeds (the begonias go in immediately, but better wait with the eggplants until three or four weeks before frost-free date). We set up the paraphernalia for germinating under lights or shift things around in the house to make room for the rough work table near a bright window. Then the seeds come. Lists are checked, seed packets against orders, and we play a sort of solitaire with the packets, grouping them according to planting time and, within these groups, according to germination medium and technique. Twice during frigid late February evenings I have become so involved in the basement with this sort of thing that I let my bedroom fire burn out. And when one is used to going to sleep with the dim glow of a banked fire to keep one company in a chilly room, that is a catastrophe.

And there's more than just seed orders: nursery catalogs with ornamental trees, shrubs, perennials and specialty flowers supply long hours of tantalizing reading; catalogs of blueberries, raspberries, strawberries, and those marvelous new Canadian strains of rhubarb bring exquisite agony because, as the man said, "There's no room . . . ." That brings about the stimulating business of redesigning the garden, which presents an opportunity to get down on paper all those dim plans that float about in summer, such as new and needed water lines, drainage for the lower garden, reduced lawn in favor of raised rockery beds (easier on the pocketbook and an aging spine), and a reduction of those fungus-prone modern roses in favor of a new bed for asparagus.

In the steamiest of houses, one can freeze oneself by sitting evening after evening watching chilling "thrillers" on television, finishing up with the invariably dismal, cold, and depressing late news. The best electric blanket can't overcome the chill of such a routine. But the delightful anticipation of garden activities is quite another thing. It brings a physical glow and certainly is mentally healthful. Who ever heard of a gardener hijacking an airliner or shooting people for kicks?

Some Special Aspects of Indoor Horticulture During Winter

Indoor gardening during the winter months can range from half-heartedly caring for a few foliage and flowering plants casually picked up at the supermarket to a serious commitment characterized by thought-out purchases of plants and special arrangements for their accommodation. Three options for winter accomodation deserve consideration if you take plant-growing seriously.

Plants under Lights

Supplying artificial light to plants is a tricky business. Daylight floods down with unbelievable intensity, more or less from all sides. No artificial light can do this. It comes down from the top. Some manufacturers have introduced lighting designs that supply some light from the side, but so far nothing comes close to the overall wash of daylight. Nor can the intensity of natural light be reproduced by artificial. Incandescent bulbs, which produce a lot of heat relative to the light they emit, will scorch the plants long before they reach solar intensity. The alternative is a great many incandescent bulbs operated

---

To Force Branches to Bloom

Bring heavily budded branches indoors; a few kinds will force as early as February 1 to 15; most will force after March 15. With a sharp knife, split cut ends 6″ (15 cm) to facilitate water uptake. Plunge branches in at least 12″ (30 cm) of tepid water, and give the tops strong light (no direct sun), cool room temperature, and humidity as high as possible. Syringing with tepid water daily helps slow sorts.
Species that can be forced include:
Abeliophyllum distichum—Korean abelia leaf
Acer rubrum—red maple
Cercis canadensis—redbud
**Chaenomeles speciosa*—flowering quince
*Cornus mas—cornelian cherry
**Forsythia x intermedia*—border Forsythia

Hamamelis japonica—Japanese witch hazel
**Lonicear standishii*—Standish honeysuckle
Magnolia stellata—star magnolia
Pieris japonica—Japanese andromeda
Prunus tomentosa—Manchu cherry
Prunus triloba—flowering almond
**Rhododendron mucronulatum*—Korean rhododendron
**Spiraea prunifolia*—bridal wreath spirea

Try, also, various fruits, such as apple, peach, apricot, and plum, and ornamental crabapples and flowering plums and peaches.
*Force easily, especially after March 1.

at a suitable distance to allow heat to dissipate before the light reaches the plants. This always has been excessively expensive; today it is prohibitive. There are alternatives.

Fluorescent lighting is relatively efficient; it gives off a lot of light with only a little heat. Most fluorescent lights supply wavelengths suitable for plant needs—largely in the visible red and to a lesser extent in the visible blue-violet areas of the spectrum. A great many plants make satisfactory to excellent vegetative growth under fluorescent light. Some flower very well under fluorescent light, others do not; but these latter can sometimes be brought to blooming size under lights and then brought into flower in a bright, cool window. There are several sorts of fluorescent light, including the tinted ones used industrially and commercially and those that produce a maximum of the wavelengths utilized by plants, also available in a variety of styles suitable for home and commercial use.

In recent years, special, very high-power incandescent lamps have appeared on the market. They look rather like photoflood lamps, and although the glare jolts human eyes, plants take it nicely. Because of the intensity of their light, the bulbs can be positioned some distance from the plants they illuminate.

I have found a bank of lights in the basement a handy thing for starting seedlings and for bringing on and maintaining house plants that are ill-suited to upstairs conditions. My set-up was inherited from an African-violet enthusiast. A 4'-x-8' (1.2m-x-2.4m) sheet of exterior-grade 7/8"- (2.2cm-) thick plywood rests on three low wooden horses. (It originally stood on eight on-end concrete blocks.) Along each side four 2' (60 cm) lengths of 3/4" (1.9 cm) galvanized pipe with flanges at each end (screwed to the plywood) serve as legs to support another sheet of 7/8" (2.2 cm) plywood. There are two more tiers like this, for a total of four flat surfaces, each 32 square feet (3 square meters). On the undersides of the top tier and the two below it, cup hooks are screwed in along the long sides 4" (10 cm) in from the edge and spaced 4" (10 cm) apart. The light comes from 40-watt fluorescent tubes held in two 1/2"- (1.3 cm-) wide industrial fixtures with a 15" (37.5 cm) piece of furnace chain at each end. Where I need high intensity, I hang the fixtures close together with the light tubes quite near the plants. For less light, I unhook and remove every other fixture and loop the chains over the cup hooks at 3" or 6" (7.5 cm or 15 cm). It isn't an especially attractive set-up, but it holds a great many plants and is flexible enough to give a variety of light intensities and "day lengths." The fixtures are grouped in batches, each batch with its own cord and plug and controlled by a household-appliance timer. I set each timer for the number of hours of illumination that best suits the plants in that area.

You can purchase all sorts of light carts, light racks, and even light units intended to be decorative enough to be considered part of the living room furniture. I find these excessively expensive and have yet to see one that would not be relegated to the basement or a storeroom in my house. In my estimation, they are not eye-pleasing any way you turn them.

Growing plants under artificial light is incompatible with energy conservation. When I plug in my basement lights the electrical meter spins and the next month's bill is exorbitant. If you are interested in saving energy, don't go in for gardening under artificial lights.

Window Greenhouses

A window greenhouse is a sort of afterthought bay window. An assortment of commercial models are offered by several companies. Most consist of a solid, opaque bottom; two glass or plastic sides; a glass or plastic top that usually slopes and in some cases opens for ventilation; and, of course, an outer face of glass or plastic. The unit, fitted with shelves inside, fastens to an ordinary sash-type window frame. From the outside there is nothing very esthetic about these contraptions, but I have seen some outstanding results on the inside produced by amateur gardeners.

In most cases, the sliding window sashes are completely removed when the window greenhouse is installed. Obviously, there is a chance of heat leakage here, and with cheaper models that is just what you get. Look for one of the designs that is double-glazed or at least glazed with very thick (double-strength) glass or thick plastic, *and* with at least 2" (5 cm) of insulated bottom (there should be a tiny drain, too, so moisture does not soak down and ruin the insulation), *and* rubber or plastic gaskets around the edge to seal off leaks when the greenhouse frame is attached to the window frame. Have a tinsmith make a watertight galvanized tray to fit exactly the bottom of the greenhouse, and coil in it an inexpensive plastic hotbed cable with built-in thermostat, set at 7°–18° C (45°–65° F), depending on your plant collection). Cover the cable with pebbles and water. Grow

plants in saucers set on the pebbles and more plants on the shelves. Because winter light is precious, the south side of the house would be the best place for one of these things.

During the day it is a good idea to pull a nearly transparent sheer nylon (or Orlon) curtain between the greenhouse and the room, so humidity stays high in the enclosure and the temperature remains cooler than the rest of the room. Open the curtain at night to protect the projecting glazed structure from frost, and on bitter nights blow room air into the unit by aiming a tiny electric fan in that direction.

The best one of these I ever saw was given over to a collection of miniature geraniums, begonias, crocuses, grape hyacinths, and other little spring things forced in squat 3" (7.5 cm) pots, and the laciest miniature ferns on the shady lowest shelf. It was charming. A window greenhouse is also an excellent winter locale for a cactus collection. The fat, moisture-storing, thorny cacti (as constrasted with the thornless "leafy" sorts) should not be watered from late fall through early spring and need night temperatures near 5° C (40° F) and day temperatures only a few degrees warmer. You will need no extra provisions for humidity; only avoid frost or warmth.

I have never tried it, but I imagine a window greenhouse would be an ideal place to start tender seedlings six or eight weeks before frost-free date. Germinate seeds inside the house at room temperature and as soon as sprouts appear move the containers to the cool, bright window greenhouse to allow stocky, vigorous seedlings to develop.

Plant Rooms

A plant room is a sort of combination sunporch and conservatory. It should be on the sunny side of the house if at all possible and equipped with French doors (or something similar) to shut it off from the heat and dry air of the house.

Not long ago I supervised the conversion of an old-fashioned, open south porch to a plant room. It involved glassing in the sides in a manner architecturally compatible with the house—no small item in the budget. Then we tore into the floor and installed some drains, brought in a couple of water lines, and mounted watertight electrical receptacles along the walls. Then we covered the floor with beautiful Spanish tiles and waterproof grouting. Along the inner wall went a cabinet which, when opened, revealed the "potting shed" complete with sink and running water, storage bins for soil, sand, and such; space for pots and fertilizers; a locked cabinet for pesticides and insecticides; and ornamental faucets for hoses and hidden hose storage cabinets at each side. There are shelves in front of some of the glass, and a center window section is enclosed in a sort of store window, 15" (37.5 cm) deep, with sliding glass panels and internal glass shelves. This enclosure is intended to house very delicate plants requiring extra high humidity. It is a sort of oversized terrarium, and requires special exhaust fans, heat controls, and humidifying equipment. In the room are recessed, very high-powered commercial fluorescent tubes and high-intensity incandescent plant lights. There is special humidifying equipment, air circulating equipment, and an exhaust system.

As you can imagine, only somebody as rich as Croesus (in this case a surgeon) could afford to build such a thing, let alone pay the monthly operating costs. But it is a dream of a room, with a tinkling fountain, comfortable garden furniture, and everything from 12'- (3.6 m-) high ficuses and giant, almost everblooming hibiscuses, gardenias, and oleanders, to tiny pileas, miniature sinningias, and dishes of *Nertera granadensis*. Bougainvillia, mandavilla, and passion-flower vines clamber about. Against my recommendation, it once housed several tiny finches and some chameleons intended to act as biological controls, but they failed utterly to eat scale insects, mealy bugs, white flies, and other ruinous pests. Finally, the livestock was moved out and badly infested plants were discarded. A weekly spray program was instituted and continues, with satisfactory control of insect pests.

Beyond such a room, one is in the realm of garden greenhouses, hotbeds, and coldframes. But that is honest-to-goodness outdoor gardening, beyond the scope of this discussion.

Plants in the Outdoor Winter Environment

There's nothing very warming about gardening outdoors from October through April in much of the North Temperate Zone. But there are certain jobs one must take care of during the fall months to get the garden ready for winter, and there are other jobs to tend to once freezing weather has begun. After that, it's largely a matter of letting well enough alone, at least in the northern half of the United States. My garden is very near the middle of the country and the weather, unpredictable most of the time, is exceptionally capricious from the autumnal equinox through the vernal equinox. During the bitter winter of 1976-77, our temperatures in late October were –8°- –10° C (14°-18° F), and later they plunged to –20° C (subzero F) for days on end. The previous winter, the first killing frost came in late November, but in 1977-78, though there was a brush of frost in low-lying areas and open spaces during the full moon of September, temperatures stayed unseasonably warm until nearly Thanksgiving. After Christmas we can always count on some quite cold weather, but it comes and goes. There is seldom a year when one cannot go out and poke up some sort of a flower in the open gardens every winter week. It may be a few "Royal Elk" English violets or some precocious forsythia flowers on low, sun-drenched branches, or something less desirable (but somehow appealing in winter), such as a dandelion or chickweed or henbit. On bitterly cold days when I am housebound I think of friends in Tucson busy with their displays of wallflower, pansy plants, and candytuft, or of other friends in the Puget Sound area, where the weather is cold and clammy, with leaden skies, but where there are cold weather vegetables and flowers aplenty. One can find the winter camellias around Mobile Bay and Houston, all sorts of tropical vegetation in southern California and Florida, and more outdoor gardening of a different sort in the San Francisco Bay area and in the mid-Atlantic states. But then I think of gardens barred to human habitation by the frigid conditions of New England, the northern tier of states, the Rockies and the Plains, and I feel better about the whole business. There's always that dandelion blooming in the roadside ditch!

When winter really comes in earnest, it is gratifying to know that everything is done that can be done to help plants in the garden make it through the chilly weather. To do these various "winterizing" chores effectively you have to think of them from a plant's standpoint. We spend the fall tightening up the house and laying in firewood—our goal is to keep warm. The real danger with hardy plants is not the chill, but chill alternating with warmth. There's no earthly way of assuring warmth to plants growing in the open garden, so we take steps to isolate them from winter warmth. But because there is nothing to do about this aspect of winterizing the garden until weather is frosty, it pays to keep busy with other projects.

Debugging the Garden

Insect pests are probably the gardener's greatest bane, followed by plant diseases. Today we can add a third name to that twosome: the Environmental Protection Agency, which has taken away many of the working tools that enable us to control insects and diseases and failed to replace them with reasonable substitutes.

Whereas some insects overwinter in an immature, dormant, and concealed fashion (egg, larval, or pupal stage), many spend the winter as adults, tucked away in various nooks and crannies. They remain in a sort of suspended animation until the warmth of spring rouses them and they multiply. One of the most useful of fall activities is to destroy the hiding places of these pests, to stir them up so they are exposed to weather that will destroy them.

The perennial border, annual beds, and the vegetable garden are among the prime hiding places of insect pests. Make it a practice to pull up all dead annuals and all vegetable plants; clip off tops of frosted perennials; rake up leaves from beds, borders, and drifts in the shrubberies and elsewhere.

Now that you have a mountain of debris, what should you do with it? In the old days, we burned it, and that, without question,

---

Ice and Snow Damage to Trees

After a snow storm, shake branches of evergreen and deciduous shrubs so that they do not get overladen with snow, which might cause breakage.

If branches become heavily ice-encrusted, spray them with a saturated solution of ni-

trate of soda. This will crack the ice, and as it melts, the solution will act as fertilizer.

If a branch should break off, use a pruning saw to even off the breaking point. Make sure the end is smooth and clean. Try to angle the wound so that water cannot collect on the open edge. The edge may be treated with "pruning paint" (asphalt varnish), though many people feel this does little good and may inhibit growth.   —E.L.

destroyed the insects and disease spores. It also smelled wonderful. On the negative side, however, there was considerable air pollution, and a large amount of valuable humus-yielding compost was lost. It is better to compost. You can look up the technique for making a compost pile in any reliable general horticultural reference. Always bear certain things in mind. A compost pile is not a trash heap that decays down slowly; such a thing is nothing more than an incubator for pests and diseases. A real compost pile heats up at least once, preferably two or three times, to a point at which insects and disease bodies are killed. You build up layers of plant waste, soil, and fertilizers (preferably both barnyard manures and chemical fertilizers), repeating the sequence until a sizable, dense, concave-topped heap is ready to ferment. Then, like the Old Testament prophet, you drench it with water to ignite the fire. And that's just what happens: copious moisture stimulates growth of fermenting microorganisms, which generate heat as they work. A properly put-together pile may actually reach the smoking point in a day or two. And by spring, the refuse should have decayed down to a form usable in the garden.

Not all your efforts in the fall garden are at the compost pile, but that's where most of them end up. Mow the lawn for the last time and compost the clippings. Clear off beds and borders and empty windowboxes and urns, and deposit the debris on the compost pile. Even late hedge trimmings and other prunings go to the compost pile, provided they are not too woody. Now the grounds are neat and clean, but beds, borders, and shrubberies are bare and exposed.

The next step is to utilize the previous year's compost, as well as leafmold made early in the fall from shredded leaves hastily put through a hot composting process. Shallowly cultivate soil in shrubberies and perennial plantings, taking care not to disturb roots. Schedule this job after three or four killing frosts to ensure that the plants are well into dormancy. Apply a generous amount of old compost or leafmold or both. Shallow-rooted ericacious plantings (rhododendrons, azaleas, pieris, kalmia, and the like) need a deep mulch of shredded leaves and leafmold. After a month of freezing weather you can scatter organic fertilizers—such as cottonseed meal, bonemeal, and manures—on top of these mulches so they wash down into the material through winter.

Turning soil is another major step in debugging the fall garden. The point of fall and winter digging is twofold: to improve the depth of topsoil and to open the soil so winter can work deeply into it, destroying exposed pests and perhaps some disease organisms, and physically and chemically modifying the soil itself. This is where techniques such as spading, double-digging, and trenching come into play. Spading is turning the top layer of soil to the depth of a spade blade, called a *spit*. Double-digging involves turning the soil more or less in place to a depth of two spits. Trenching means turning soil to a depth of two or more spits and usually bringing poorer soil to the top for improvement while burying well-worked-up topsoil. Entire books have been written on these techniques, every gardener ought to be familiar with all three and to practice them when possible to increase the depth of fertile, well-drained, well-aerated topsoil.

But our topic is debugging, not soil improvement, so we remind ourselves to leave the surfaces of spaded vegetable patches and flower beds in coarse clumps—never raked down or otherwise smoothed—because we want frost, snow, and ice to penetrate deeply between these clods. More compost and manures will be dug into the soil as we work it. Spade any time the soil is neither too wet nor frozen.

Mulching and Covering

By the time you have cleared your property of weeds, dead annuals and vegetables, perennial clippings, lawn clippings, and fallen leaves; have mown the lawn, cleaned, cultivated, and mulched perennial beds and shrubberies and turned the soil on all empty ground, building and turning compost piles between times, surely frost is beginning to bite into the soil. When the soil is frosty, debugging (which also includes destroying infections of fungi and bacteria) becomes secondary; it is time to get on with other aspects of winterizing the garden.

Pruning, it happens, is not one of them, although many people mistakenly believe a fall pruning is beneficial. Pruning tends to stimulate growth activity. When plants are pruned in fall or early winter, they may break into precocious growth, especially if weather is mild. Dormant (winter) pruning is a February-March proposition, and late spring/early summer pruning—April-June, depending on latitude—is probably best of all.

Covering plant roots to prevent freezing and thawing *is* an important operation. Several different methods can be used in most gardens.

Where snow cover is deep and persistent, very few plants derive any benefit from winter covering. Where winters are very cold and snow cover is scant or comes and goes, some sort of covering is advisable on most herbaceous (nonwoody) plants that keep green leaves through winter. Any sort of tight or wet covering will damage these, so compost, leaves, and similar mulches are out of the question. Loose, fluffy straw is a possibility, but it will compact and go soggy under a wet snow. Evergreen boughs are the best bet. Since this sort of covering usually is not put down until the end of the year, wait and gather up old Christmas trees; cut off the branches, and use them to protect rosettes of shasta daisies, foxgloves, Canterbury bells, and the like. In areas with early spring weather that wavers between warm and cold, it is a good idea to spread a blanket of branches over spring-flowering bulb beds to keep the frost in the soil until the worst of winter is past. For perennials that go completely dormant, almost any mulch will do, although such species as the irises, most artemisias (in fact, most plants in the herb category), and alpine and rock-garden plants suffer more from a humus-yielding cover than from being left uncovered.

Some rock-garden and alpine species are often fairly shallow-rooted, though the most common ones are not. But because heaving will ruin the shallow-rooted species, many growers make it a practice to add additional scree material, such as stone chips, to the rock garden in midwinter to keep freezing and thawing to a minimum. *Heaving* refers to the upward-lifting action of soil caused by alternate freezing and thawing of soil. Shallow-rooted plants may be hoisted completely out of the soil, where they desiccate and perish. High-mountain species, which depend on a permanent winter snow cover to shelter them, do best under a deep, loose blanket of pine branches that hold their needles well. Furry-leaved sorts, intolerant of winter moisture, can be effectively sheltered by a flat pane of glass or plastic supported by slotted stakes at the four corners. This keeps the crown of the plant dry while giving it the bright light and air it needs to survive.

Roses are *the* big problem. They wouldn't be if rose breeders paid half as much attention to cold-hardy and diseaseproof breeding stock as they do to commercial promotion programs, but there it is.

Most of our modern hybrid tea, floribunda, polyantha, and grandiflora roses are not reliably hardy. Where there is a question about frost tolerance, stop fertilizing soon after midsummer so rose plants make harder, less succulent tissues; reduce artificial watering; maintain your pest- and disease-spraying program so plants are as healthy as possible. When frost has blackened the leaves, clean up the rose garden. Clip away buds that will fail due to cold. If you live in a place where winters are windy, either loosely bunch and tie the canes or, better, trim all the plants to a uniform height—not too low or too early, since pruning stimulates growth; about 30" (75 cm) should do it. Then, when there has been a skim of frost in the soil, bring in soil or pull it from the middle of the spacing between plants—take care not to expose roots—to cover the base of each plant with a broad cone a foot or more high. It is important to realize that fewer roses are lost because they have been left uncovered than die because they have been covered too early. If you plan to mound your roses, wait until they are truly dormant.

Deep, long-lasting snow is a blessing to gardeners. Over much of the deep snow belt the soil below is so insulated that it barely freezes, even though the temperature above the snow plunges to below zero for long periods. Snow brings atmospheric nitrogen into the soil, thus acting as a fertilizing agent. But it also weighs down willowy plants. Columnar evergreens, especially multiple-stemmed ones, may be sprung apart, their form ruined. The old-fashioned way to cope with this sort of problem was to rush out in late fall and wind wide strips of rag or burlap around the plants in a fairly tight spiral to keep them narrowly upright. A better way, and one that serves the year 'round, is to use the rag strips to tie stems together (make figure-eight ties so branches are not girdled) inside the shrubs at several levels. This holds the plants through the worst weather without making the garden look as if it were girding up for the next Ice Age.

People used to greatly overdo this protection business, to keep the hired gardener busy through the winter, I suppose. Christmas trees were stuck in rhododendron beds to reduce sunscald of foliage or were used to cover prize specimens. It is far simpler to plant varieties that are not susceptible to scald. Snow fences used to be a favorite gimmick in gardens; they were wrapped in cylinders around hydrangeas and tree peonies and filled with loose leaves or straw, making the garden look like a barrel-salesman's yard. By taking advantage of modern species, varieties, and cultivars, we can achieve delightful

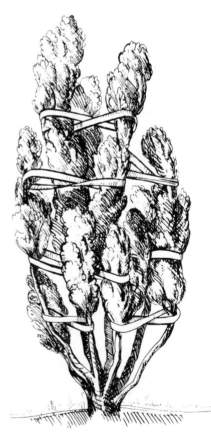

Shrub with figure-eight ties

landscapes that require almost no winterproofing at all. It's really the better way.

Today the gardening picture is shifting fairly rapidly; hired gardeners faded from the scene during the Depression years and homeowners became their own gardeners. Gardening became simpler; and a lot of it was lost, too; America is the poorer because towns, especially, have lost much of their horticultural beauty. But amateur gardening has become a stable pastime for many people, and many of us tend to make up for a lack of knowledge with a plethora of garden gadgets. And they are mostly power driven. Riding mowers, power rakes, power clippers, edgers, tillers, sprayers—you name it and it uses electricity or gasoline.

But we face a power curtailment today that appears to be permanent. Let's get rid of the power gadgets, and the parts of the garden that demand them. Power cultivating has always been a questionable procedure because most powered soil machines mill the soil to powder, which quickly settles, or is beaten by rain to an airless, poorly draining, compact medium, ill-suited for plants. We can do without powered lawn gadgets if we reduce grassy areas to a bare minimum. A velvety greensward eats up more of the gardening budget than any other garden feature; fertilizers, weed-control chemicals, pesticides and fungicides, power mowers, edgers, aerators, and sweepers—it all adds up. A tiny gem of a lawn can be managed by hand; you'll appreciate the exercise and can forego those expensive and energy-consuming trips to the gymnasium or indoor tennis club. The extra space can go to hand-managed shrubberies and productive vegetable gardens, cutting gardens, and the like. All the same, I will fight to the end to keep my power sprayer (mostly for the sake of the semidwarf fruit trees, though my rose gardens, too, come into this picture) and I would like to be able to afford to use my power compost shredder through the fall months.

I figure it this way: Through late fall and early winter, when the low thermostat setting makes the house seem too clammy, I can always go to the garden and work up a good sweat doing something active and beneficial. And, thus far, nobody has figured out how to tax these efforts. When all else fails, there is always wood to saw and split. Abraham Lincoln pointed out that supplying your own firewood warms you twice: when you cut and split it, and when you burn it. It's true. Everybody has the option of sitting around growing chillier or stirring about to keep warm. The former may be more popular, but I personally find the latter far more gratifying.

Cars and Winter Driving

You wouldn't go out on a cold winter day dressed in summer clothing, and you can't expect your car to operate well in winter without preparation appropriate to the season. Certainly winter driving conditions require greater precaution, but there is more to winterizing a car than scraping ice off the windshield. There are things to be done in advance to insure that your car starts and runs well throughout the snowy season, no matter how cold, slushy, or icy it may be outside.

Cold weather puts a greater strain on your car than warm or even hot weather does. Oil thickens in the cold, creating a drag that the starter must struggle to overcome. The starter is effective only if the battery is strong; and a battery can deliver only 65 percent of its power at $0°C$ ($32°F$) and about 40 percent at $-15°C$ ($5°F$). Gasoline must be vaporized before it will combust; extreme cold makes this difficult, so the engine struggles until it warms up. Furthermore, the engine's cooling system, which you might think has its biggest job in summer, actually does an equally important—though different—job in winter, and faces especially difficult operating conditions. First, it must operate in subfreezing temperatures, which means it must not only contain antifreeze, but the right amount to keep the cooling fluid flowing. Second, it must control heat flow so the engine warms up as quickly as possible, and then it must provide the heat to warm the passenger compartment. These functions require a clean system with properly operating mechanical components.

You want to be able to keep moving once you've got your car started and running warm. To minimize the chance of getting stuck in snow or ice, you need the right tires for winter driving. If you encounter severe conditions, you want to have what's needed to get the car going again—and you back inside it.

Keeping warm in your car this winter has many facets. Let's look at them in turn.

Winterizing Your Car

Whether you plan to drop your car off at a local gas station or do a substantial portion of the work yourself, you should know which basic tasks are part of the winterizing process. You will probably get more for your money if you know what you're talking about when you take your car in for service, and you certainly need to know what's involved before you don your coveralls and seize your wrenches.

Basically, winterizing should include a complete tune-up; oil change; service to the cooling system, including heater and defroster; battery service; and a general maintenance check of the body, including windshield-cleaning system, lights, and tires.

Tune Up

A well-tuned engine will not only give better fuel economy, but is also part of the vital insurance you need for keeping your car running well in winter. You should have your points and plugs checked, and replaced if necessary. The timing should also be checked, it may need adjustment. At the same time, the automatic choke and exhaust systems must be in top working order, so have them looked at as well. Ask your mechanic to check the ignition system: Are all wires in good condition? Is the distributor cap intact? No matter what shape the rest of the car is in, if you can't start it, you won't be going anywhere.

Oil

Oil is an important factor in an engine's winter preparedness. If the oil is dirty or too thick, it can't lubricate the engine in cold weather.

Oil also plays an important role in the engine's starting ability. It takes power for the starter motor to spin the crankshaft through a pan full of cold, thick oil. If the drag of the oil is great, the engine will crank sluggishly. There is no hard-and-fast rule about how frequently

---

Doing It Yourself: Repair and Maintenance Manuals

In the pages that follow, we are going to be talking about what constitutes a winterized car. If you like, you can drive on up to your local service station and ask for the works (you may get just that, though not what you had in mind), or you can get specific and ask for a check of various systems as well as a flush of this and a liter of that. Or, if you are in the mood to get more intimate with your car, you can do some or all of the work yourself.

That's what the Doing It Yourself Projects are about. If you follow the instructions carefully, you will be able to flush out your own cooling system, change the oil and oil filter, clean up the battery terminals, and a handful of other essential winterizing tasks. Bear in mind that these instructions are general; they describe what to do in most cars. Your car may have a radiator hose that looks slightly different or is in a slightly different position than the radiator hose in another model; the capacity of your car's cooling system will not be the same as that of a larger or smaller, older or newer car that someone else might have. So in addition to the instructions herein, you will need a repair and maintenance manual of some sort.

Repair and maintenance manuals come in two basic forms. One is a large, general book about the way cars work and what needs to be done to them to keep them in good working order. The other is a manual that assumes you know a lot about cars already, and concentrates on the peculiarities of a particular make, model, or year of car.

The large general books cover the range from primers for people who know absolutely nothing about cars to textbooks and workbooks in auto mechanics. Some of them are written in plain language, whereas

others assume that you have at least heard of carburetors, manifolds, alternators, and other parts of a car's anatomy. A sampling includes:

Chilton's Auto Repair Manual: 1978. $13.95. Chilton also publishes a comparable manual for imports, $18.95.

How to Service and Repair Your Own Car, by Richard Day. Harper & Row, 1973, $12.95.

The Auto Repair Book, by John Doyle. Doubleday, 1977, $12.95.

The Complete Do-It-Yourself Handbook for Auto Maintenance with the Repair-O-Matic Guide, by Ronald V. Fodor. Prentice-Hall, 1977, $12.95.

The Complete Book of Car Maintenance and Repair, by John D. Hirsch. Scribners, 1977, $12.95.

Car Care, by Mechanix Illustrated Editors. Arco, 1968, $4.95.

Petersen's Big Book of Auto Repair, by the Petersen's Publishing Company Staff. $11.95.

Auto Repair for Dummies, by Deanna Solar. McGraw-Hill, 1976, $12.95 (hardcover), $7.95 (paperback).

The Time-Life Book of the Family Car, Time Inc., 1973, $14.95 (hardcover); Rand McNally, 1975, $7.95 (paperback).

The manuals that are devoted entirely to a specific make, model, and/or year are too numerous to list here. Several publishers concentrate on providing an up-to-date list of books of this type, revising old editions and bringing out new ones each year. Foremost among these publishers are Audel,

Chilton, Clymer, and Hearst. Audel's *Can-Do Tune Up* series covers the cars of each domestic manufacturer and can be bought with a cassette ($7.95 with, $4.95 without); Chilton issues model-by-model volumes for all domestic cars; Clymer publishes books on imports (by manufacturer; some are $7, some $9); and Chilton has an extensive list of manuals for older cars (even from the early 1960s, from $10.95 to $18, depending on the subject). Between them, just about every car still on the road is covered. There are enough books about Volkswagens— such as *How To Keep Your Volkswagen Alive,* by John Muir and Gregg Tost, John Muir Productions, 1975, $7.50—to fill a shelf.

The most specific manuals of all are available through the parts departments of automobile dealers. These are the repair and servicing manuals that are published by the car makers for use by mechanics in their service departments. They are expensive, between $20 and $30. Because they are written for professional mechanics, they make no concessions to the layperson in language or explanations. They are valuable because they describe procedures that are specific for a given car, and refer to the manufacturer's parts by their original number.

—Alan Ravage

---

engine oil should be changed. In general, though, late-model cars running on unleaded gasoline should have their oil and oil filter changed every 9600 kilometers (6000 miles) or six months, whichever comes first; older cars and those using leaded gas should have theirs changed every 6400 km (4000 mi) or three months; diesel-engine cars should have the change even more frequently. Your best guide is your owner's manual, as modified by your personal experience with your car.

Whether you have the oil installed at a service station or do it yourself, you should know how to choose the right oil for your engine.

All oil cans are marked according to a grading system of the American Petroleum Institute, but there is no policing of the refiners, so you have to rely on reputation. If in doubt buy a name brand you know.

Motor oils are labeled according to thickness—the higher the number, the thicker the oil; the thicker the oil, the less easily it flows at moderate temperatures. Modern oils contain additive, however, which produce the free-flowing qualities of a thin lubricant at low temperatures and the resistance to breakdown of a thick oil at high temperatures. For example, 10W-40 on the can means that the oil flows as freely as a 10-weight oil at -17°C (0°F) but has the thickness of a 40-weight oil at 99°C (210°F). These are respectively the stardard temperatures for winter starting (that's what the W stands for) and optimum running set by the Society of Automotive Engineers.

The virtue of double-weight oils is that the oil is thin enough when cold to not impede starting and flow easily to all the moving parts, but when the engine (and the oil) is hot, it is sufficiently thick so its lubricating film won't break down.

A single-grade oil, 10 for instance, would be fine for winter starting, but probably would be too thin at high temperatures. A single 40-weight oil would be thick enough at high running temperatures, but probably would be much too thick at low starting temperatures. It would be very difficult to start the engine, and the oil would probably be moving so slowly through lubrication passages that the moving parts would suffer until the engine warmed up.

A 10W-30, 10W-40, 10W-50, or 10W-60 oil is an excellent choice for most winter use, with 10W-50 the best if you do turnpike driving. If the winters are unusually cold, with consistently below-zero temperatures, a 5W-20, 5W-30, or 5W-40 oil is advisable. Check your owner's manual for advice on which to choose for your car.

If you can use one of the 10W oils, and plan to do your own oil changes, consider saving money by buying a case of twenty-four liter (quart) cans. This oil is suitable for use the year round, so you won't be buying an outrageous amount of oil. The typical American car engine requires 3.8 liters (4 quarts) for an oil change, plus an extra liter (quart) when the oil filter is replaced.

Another possibility is a synthetic oil, a more expensive initial purchase but a money saver in the long run. Synthetic oils are very thin at low temperatures, so they flow easily and make starting simple. They have excellent resistance to thin-out, making them equally effective at high temperatures. They reduce friction so well that gas mileage is frequently improved. The oil also has long-life properties; you can safely leave it in for a few years if you change the oil filter annually and top up with the same brand of synthetic whenever the dipstick reads low. The major hitch is price—$4 to $5 a liter (quart), but it pays for itself in the long run.

Doing It Yourself: Changing the Oil

You Will Need:

wrenches
oil
oil filter
flat pan (a plastic 5.6-liter [6-quart] household dishpan will do fine)
funnel
drive-on ramps (optional)

Two wrenches are needed—one to remove the filter, the other to work the oil drain plug. A standard oil-filter wrench can be used on a range of filter sizes, but some filters are considerably larger or smaller than standard, so be sure the oil-filter wrench you have will fit your car's oil filter.

To check the fit of an oil filter to the wrench, insert the filter in the wrench's circular band and move the wrench handle to close the band. If the band closes tightly around the filter can, you've got the right size.

If you have a set of wrenches for household use, you may have one that will fit the oil drain plug on your car. If you don't, and do not want to invest in a complete set of wrenches, you can buy a combination filter-and-drain-plug wrench. The combination wrench is generally available in discount and auto-supply stores for about $5.

You should be able to find the correct oil filter for your car in a catalog, and buy it in a discount store for the lowest possible price. If not, shop at one of the auto-parts stores that serve the professional mechanic. The discount won't be quite as good as that given the professional, but the salespeople are usually knowledgeable and used to helping weekend mechanics. You're an important part of their business, so don't be embarrassed to ask questions.

If you buy at an auto-parts store, you'll get a name-brand filter almost all the time. If you buy at a discount house, you may find a choice between a name brand and a discount-house brand. The house brand probably isn't a bad filter, but it generally isn't as high quality as the name brand. The name brand may cost a dollar more, but it's probably worth it.

You may have to invest in equipment to raise the front end of your car off the ground. To find out if this is necessary, slip under the car and, using a flashlight, see if you can reach the oil drain plug. It looks like a bolt threaded into the lowest point on the oil pan, at the bottom of the engine. Also look for the oil filter, a cylindrical cannister protruding from the engine (it may be easier to see from the top of the engine compartment on some cars). If you can't quite reach either or both, you'll need to raise the car. The simplest and least expensive method involves a pair of drive-on ramps, sold in the automotive departments of discount stores for $16 to $20. You position them in front of the car, drive up onto them, apply the emergency brake, and put the car in park (if you have automatic transmission) or first gear (if you have manual transmission).

Doing the Work

Don't wait until winter to change your oil. Unless you have a luxuriously heated garage to work in, it will be cold going. Pick a mild fall day. Nor should you work on a cold engine; the oil will not drain thoroughly when it is at its thickest. It is best to run the engine until it's warm

(not hot), or let it cool down a bit from running temperature.

If necessary, raise the car on ramps, then slip underneath with the wrenches. Have the flat pan handy to catch the oil.
1. Drain the Oil
   Place the appropriate wrench on the oil drain plug and loosen by pulling in a counterclockwise direction. If the drain plug is very tight, whack the end of the wrench shank with your palm or tap it with a hammer. Once the plug is loose enough to turn by hand, put down the wrench. Position the catch pan underneath and unthread the plug with your fingers. When the plug is hanging on its last thread, twist and quickly move the plug and your hand away, or oil will pour out onto your arm.

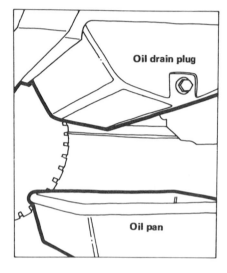

Oil drain plug

Oil pan

Allow several minutes for all the oil to drain out, then carefully thread the plug back in by hand (turning clockwise). The threads may not catch easily; if there is any binding, don't force the plug or you'll damage the threads. Once the plug is threaded hand tight, secure it with the wrench, but do not exert unusual force.

2. Change the Oil Filter
   On most cars the oil filter is quite accessible from either the top of the engine compartment or underneath. On a few cars the filter is mounted low and seems to be surrounded by other parts of the engine. Your owner's manual will tell you exactly where to find the oil filter on your car.

   Use the oil-filter wrench to loosen the filter cannister, then unthread it the rest of the way by hand. Keep the catch pan handy if the filter is mounted horizontally or at an angle, for some oil will probably spill out of it.

   Once the old filter is off, you're ready to install the new one. Find the rubber O-ring on the top of the filter and smear it with clean engine oil. Then thread the filter onto the engine by hand. If some oil has gotten on your hands and/or the exterior of the filter cannister, wipe dry with a rag soaked in cleaning solvent, then complete the job by hand-tightening the filter. It is not necessary to use the wrench to tighten it; the filter O-ring will heat seal to the engine.

3. Install the Oil
   Remove the oil-filler cap, located at the top of the engine (check your owner's manual if you are not sure where it is). Insert a funnel in the cap opening, puncture the oil can, and start pouring in the oil. Stop after you've poured in 2.8 liters (3 quarts) and pull the oil dipstick. Wipe it clean, reinstall all the way, and withdraw it once more. Read the oil level on it. Add oil as required to bring the level up to the dipstick's full mark.

   Note: if the car is up on ramps, the angle will produce a false reading on the dipstick. Stop after 2.8 liters (3 quarts) of oil and do the final top-up when the front of the car is on the ground.

   Start the engine and immediately look underneath for leaks. If the engine does not leak after a minute of idling, stop it, lower the car to the ground if necessary, and recheck the oil level. The oil filter will have filled up with oil and you'll have to pour in some more to bring up the dipstick level.

   If you do see an oil leak, determine whether it's coming from the oil drain plug or the filter, the only two items you touched. Stop the engine and tighten the plug or the filter as required, then restart the engine and recheck.

You Will Need:
tee-kit
razor blade
garden hose
antifreeze
Fast Flush (optional)

There are many ways to flush out the cooling system. For the beginner, the simplest way is to use a flushing tee-kit, available in most auto-supply stores. Priced from $2 to $3, the kit comes in three sizes. Use the one that fits the inner diameter of the heater inlet hose in your car—1.3 cm, 1.6 cm, or 1.9 cm (1/2", 5/8", or 3/4").

Included in the kit are a tee-connector, which fits into the heater hose and thereby gives you access to the cooling system; a pair of hose clamps to hold the tee-connector in place; a double-female connector, which is used to connect a garden hose to the tee-stem; a cap for the tee-stem, so it can be sealed when it is not in use; and a water deflector, which fits into the radiator cap neck.

Buy enough antifreeze to equal at least half the fluid capacity of your car's cooling system, and preferably 70 percent. That is, if the capacity is 15 liters (16 quarts), you need at least 7.5 liters (8 quarts), but 10.5 liters (11 quarts) would be the ideal.

To determine the capacity of your car's cooling system, check the specifications section of your owner's manual.

Doing the Work

To install the tee, find a convenient working location anywhere along the length of the inlet hose, and cut it in two with a razor blade. Slip the hose clamps in the kit over the cut ends of the hose and slide them out of the way. Push the tee-necks into the cut ends of the hose as far as they will go, then move each clamp back to the midpoint of its joint with the tee-neck. Tighten the clamps, and the tee is installed. This is a permanent installation.

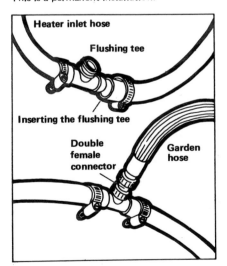

Heater inlet hose

Flushing tee

Inserting the flushing tee

Double female connector

Garden hose

Flushing the cooling system

The next step is to remove the radiator cap. Press your palm down on the cap, twist counterclockwise until it hits a stop, then press down a bit harder and continue twisting counterclockwise until the cap hits a second stop. At that point, twist

slightly and lift it up and off. If the cap has a lever in the top, lift the lever before you begin. You cannot use your palm to turn this type of cap, but with the lever up it shouldn't be difficult to twist off.

Look at the coolant through the cap opening in the radiator. If it's dirty and rusty, obtain a can of nonacid Fast Flush, a cleaning agent available from auto-supply and discount stores. Pour it into the radiator, replace the cap, and drive the car about 160 kilometers (100 miles) to give the flushing agent a chance to work. Now you're ready to clean out the system.

Begin by pressing the rounded end of the water deflector into the radiator cap neck, pushing down until it seats in place. Aim the deflector so the nozzle faces forward.

Remove the tee cap, thread on the double-female connector and the garden hose, making the connections as tight as you can by hand.

Start the engine and turn the heater control to warm or maximum. Turn on the water to the garden hose. The water will flow into the heater hose and from there will circulate through the heater core and the engine, to the water pump and into the lower radiator hose, up the radiator and out the water deflector. (The deflector aims the water forward so it doesn't splash on the spinning fan.) The course of flow is counter to normal circulation in the engine and radiator, so it helps dislodge rust particles and dirt, flushing them out of the cooling system.

When the water running out of the deflector is clean for a period of about two minutes, stop the engine, shut off the water to the garden hose, disconnect the double-female connector, and remove the water deflector.

Note: The car's heater/defroster is an extension of the cooling system. If the heater has not been performing well, the problem is most likely in the thermostat or a heater core clogged by rust. Before proceeding to the next step (adding the antifreeze), you should determine what is the matter with the heater and correct the problem. The thermostat should be tested and replaced if necessary, and the heater core should be flushed (see p. 151). Do not add antifreeze until these jobs have been done, or it's wasted energy and antifreeze all around.

Before you start pouring in the antifreeze, make room by draining some water from the cooling system. The radiator is the most convenient drain, for in most cases there is a cock or drain plug at the bottom of the radiator. The cock can be twisted open with a pair of pliers; the drain plug must be removed with a wrench of the correct size.

Drain plugs or cocks were not installed in the radiators of all water-cooled Volkswagens (Rabbit, Scirocco, and Dasher), the Audi, and many 1972-73 General Motors cars (Chevrolet, Pontiac, Buick, and Oldsmobile). On these cars you must disconnect the lower hose from the radiator to drain it. Check your owner's manual or a repair guide for the best way to do this.

Once the radiator has drained, close the drain cock (or refit the plug and tighten with the wrench, or reinstall the radiator hose and tighten the clamp). Now you're ready to pour in the antifreeze. Because the radiator holds only about 25 percent of the cooling-system capacity, you won't be able to pour the entire quantity of antifreeze into the radiator. When the radiator is almost full, start the engine and turn the heater to warm or maximum. Water will

start to flow out of the opening in the flushing tee (remember, the cap is still off). As the water in the system pours out of the tee, the level in the radiator will drop. Just keep pouring antifreeze into the radiator as quickly as possible; when you have it all in, quickly thread the cap onto the tee, tighten it, and turn off the engine.

If you have an overflow reservoir on your car, you may be able to avoid this somewhat tricky procedure. The reservoir is usually large enough to hold the remaining antifreeze. Just empty the reservoir and pour the antifreeze in. The reservoir on some cars is held in place by screws or bolts; if they are hard to remove, you may not be able to get the reservoir out to dump the contents. The solution is to empty the reservoir with a siphon. Although it is desirable to clean out the reservoir with water before filling it with antifreeze, the coolant inside will not be overly rusty if you have given your system periodic care.

There is one other limitation—synthetics are not always suitable for older cars. The oil is very thin (equivalent of a 5W-20), and in a worn engine it may leak out or slip past piston rings and valves into the combustion chamber and burn. This is also a problem for conventional oil, but conventional 5W oils are normally limited to winter weather in cold areas. Synthetics, however, are intended for year-round use.

Cooling System

The cooling system's basic job is to transmit heat from the engine, which burns a mixture of gasoline droplets and air. Temperatures in the cylinders can reach thousands of degrees, even when it's well below freezing outside; without the cooling system, the engine would not last more than minutes.

A mixture of water and antifreeze (called the coolant) is circulated through passages in the engine by the water pump. The coolant absorbs heat from the cylinders and, when it's warm enough, it causes a temperature-sensitive valve—the thermostat—to open. With the thermostat open, the coolant can flow through a hose into the radiator, where it gives up its heat to the surrounding air in the radiator tubes. A fan on the engine draws air through the radiator (between the tubes) to promote the heat transfer from the coolant. After the coolant has given up heat, it is drawn back into the engine by the water pump and recirculated.

Many late-model cars have a radiator-overflow reservoir connected by a hose from the top of the radiator. As the coolant gets hot, it expands, building up pressure throughout the cooling system. Within limits, this is desirable, for the pressure build-up prevents the coolant from boiling in summer. However, the components of the system—particularly the radiator and hoses—can withstand only a certain amount of pressure. When pressure from the heated coolant exceeds that level, it pushes open a valve in the radiator cap, and

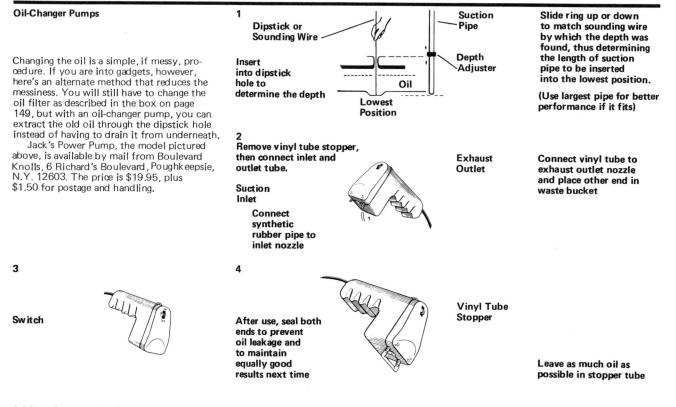

Oil-Changer Pumps

Changing the oil is a simple, if messy, procedure. If you are into gadgets, however, here's an alternate method that reduces the messiness. You will still have to change the oil filter as described in the box on page 149, but with an oil-changer pump, you can extract the old oil through the dipstick hole instead of having to drain it from underneath.

Jack's Power Pump, the model pictured above, is available by mail from Boulevard Knolls, 6 Richard's Boulevard, Poughkeepsie, N.Y. 12603. The price is $19.95, plus $1.50 for postage and handling.

1
Dipstick or Sounding Wire

Insert into dipstick hole to determine the depth

Suction Pipe

Depth Adjuster

Oil

Lowest Position

Slide ring up or down to match sounding wire by which the depth was found, thus determining the length of suction pipe to be inserted into the lowest position.

(Use largest pipe for better performance if it fits)

2
Remove vinyl tube stopper, then connect inlet and outlet tube.

Suction Inlet

Connect synthetic rubber pipe to inlet nozzle

Exhaust Outlet

Connect vinyl tube to exhaust outlet nozzle and place other end in waste bucket

3

Switch

4

After use, seal both ends to prevent oil leakage and to maintain equally good results next time

Vinyl Tube Stopper

Leave as much oil as possible in stopper tube

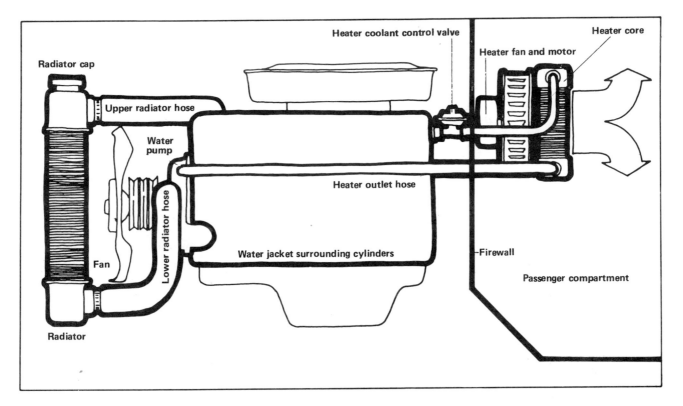

Labels in diagram:
- Radiator cap
- Upper radiator hose
- Water pump
- Lower radiator hose
- Fan
- Radiator
- Water jacket surrounding cylinders
- Heater coolant control valve
- Heater outlet hose
- Heater fan and motor
- Heater core
- Firewall
- Passenger compartment

Simplified side view of cooling system and heater

coolant flows through the hose into the overflow reservoir. When the engine cools down, the coolant in the system contracts and creates a vacuum, which sucks open a second valve in the radiator cap. Coolant drawn from the reservoir flows through it, automatically topping up the radiator once more.

On cars without this reservoir, the coolant simply flows through the hose to the ground and is lost. This is one reason that any cooling system without a reservoir must periodically be topped up with coolant.

The hot coolant is a convenient source of heat, and it is used to provide warmth for the passenger compartment. A component called the heater core, which looks like a miniature radiator, is connected by two hoses to the engine.

The heater core is located in a duct with passages under the dashboard. Next to the core is an electric fan, called the heater blower motor. When you turn it on, it blows air through the heater core, in much the same way a fan placed in back of a home radiator would. As the air is blown through the spaces between the heater-core tubing, it is heated and flows through the duct into the passenger compartment, where it also may be diverted to defrost the windshield.

The heater can provide heat only when the coolant is warm, which is why your heater doesn't function for the first few minutes of driving in cold weather.

More than 80 percent of the antifreeze installed in American motor vehicles is poured in by owners. In most cases, however, what owners do is open the radiator drain cock (or remove the drain plug), let the radiator drain, then close the cock and fill the radiator with antifreeze. Unfortunately, on virtually all cars, draining the radiator leaves 60 to 75 percent of the old mixture in the engine and heater.

To insure that the cooling system functions properly in winter—and that the heating system delivers warmth—you should flush out the old coolant and refill with a fresh mixture of antifreeze and water. This should be done every two years on newer cars, once a year on

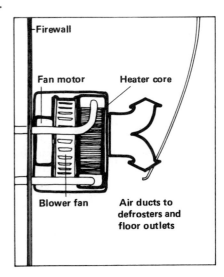

Labels in diagram:
- Firewall
- Fan motor
- Heater core
- Blower fan
- Air ducts to defrosters and floor outlets

Heater box

cars five years or older. If you don't, rust accumulations will clog cooling-system passages, particularly the heater-core tubing, which is extremely narrow.

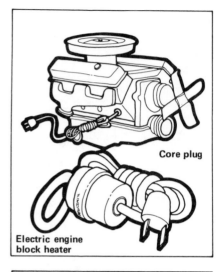

Core plug

Electric engine block heater

Heater Trouble

If your car heater is not doing the job of heating the interior or if it is not working at all, the cause is probably one of the following: inoperative heater blower motor; clogged heater core; defective cooling-system thermostat; defective heater water-control valve.

If it's the blower motor you will know it right away because the heater will not operate at all—you simply will not hear the blower when you turn on the heat. The problem could be as simple as a blown fuse. Your owner's manual or a repair manual for your make and model car will tell you where to find the fuse that corresponds to the heater blower motor, and how to check if it's intact. If the fuse is good and the blower still does not work, take the car to a mechanic for further examination.

If the cooling system has not been flushed before, the cause (or at least a contributing cause) for poor heater performance could be rust and dirt in the heater core. You may be able to clean out the heater with a maintenance procedure called reverse-flushing (see Box 5), which means running water through the heater core in a direction opposite that of normal circulation.

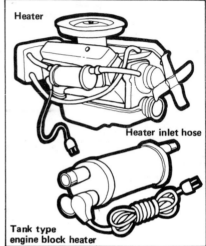

Heater

Heater inlet hose

Tank type engine block heater

The coolant thermostat is normally a reliable device. When it malfunctions, the problem is frequently that it opens too soon and too wide in winter. The result is that the engine coolant never reaches ideal operating temperature—(82°- 96°C (180°- 205°F)depending on car)—but stays in the 54°- 71°C (130°- 160°F) range. When the coolant temperature is this low, the heater performance drops enormously. You should have a suspect thermostat tested; if it is found to be defective, replace it.

Many cars, particularly those with air conditioning, are equipped with automatic coolant control valves for the heater. When you turn the heater on, you automatically activate the coolant control valve, which opens to allow coolant to flow from the engine into the heater core. If this valve doesn't open, the heater won't work.

The valve is operated by vacuum from the engine, via a thin hose attached to it. To check the valve, just pull off the hose, turn on the engine and let it warm up, and set the heater control to warm or maximum. Put a hand on the heater hose, somewhere between the valve and the back of the engine compartment. It should be cool.

Reconnect the vacuum hose to the neck on the valve; the hose should become warm very quickly. If it doesn't, the valve is stuck closed and should be replaced. If you are so inclined, you can do the replacement yourself; otherwise, bring the car to a garage.

Electric Coolant Heaters

Even if you have kept your heater and cooling system in tiptop shape and have tracked down and solved any of the problems just described, your heater may not do all you wish it to in very cold weather. There may not be anything wrong with the heater; it's just that the car heater cannot warm the passenger compartment until the coolant is hot. On a cold winter day, this can take five minutes or more; if you're taking a short trip, you may not get any heat.

There's an inexpensive cure for the short-trip problem, and it will also help you protect your engine from the ravages of cold weather. Electric coolant heaters, which plug into household current, are available from about $7 up to about $30, plus installation.

There are many types of coolant heaters, but the most widely available, and the one that can be installed on virtually any car, has a small tank mounted somewhere in the engine compartment and connected to the heater's inlet hose and the engine block. It is sold in a kit with all necessary clamps and fittings, and it is available in many different wattage ratings. To choose the right size for your car, consider the following questions:

Do you need fast warmup or will the heater be plugged in overnight? If it can be plugged in overnight, a smaller wattage unit will do. Also, if you remember to plug the heater in as soon as you arrive at your destination, while the engine is still warm, all the heater needs to do is maintain coolant temperature, not heat up cold coolant.

What size engine do you have? A small four-cylinder engine doesn't need as big a heater as a V-8. If you have a four-cylinder and can plug the heater in overnight, a 400- to 500-watt heater will do. For extra capacity in severe winters, or for fast warmup, an 850-watt is a practical maximum. If you plug in while the coolant is still warm, however, 500 watts should do even in severely cold weather.

If you have a six-cylinder, a 500-watt is adequate for overnight use, if you plug it in while the engine is warm. Otherwise, choose one with a 750- to 850-watt rating. For fast warmup or extra capacity in very severe weather, a 1000-watt heater should do the job.

For a V-8, an 850-watt heater is the minimum; choose a 1000-watt if you want an extra measure of coverage in very cold weather. For very fast warmup, a 1250- to 1500-watt heater will be necessary.

Whatever wattage you choose, look for a design that includes a thermostatic control to shut off the current when coolant temperature is at the intended level, usually 71°- 82°C (160°- 180°F). The thermostatic control adds to the cost of the heater, but in the long run saves you money on electricity.

Another possibility is an engine block heater. It is not recom-

---

Doing It Yourself: Reverse Flushing the Heater

The heater has an inlet hose (the one that goes to the top of the engine) and an outlet hose (the one to the water pump). With the tee in place, you are flushing the radiator and engine in a direction opposite that of normal circulation, but the water is going through the heater in the *same* direction as it normally does.

Reverse flushing the heater is a simple procedure, and it should be done after flushing the cooling system and before adding a fresh fill of antifreeze. Here's how:

Disconnect both heater hoses. The hose to the engine is usually connected to a water valve very close to the engine. To flush the heater, you should disconnect the hose at the valve, not between valve and engine.

The normal coolant circulation is from the inlet hose, through the heater core, to the outlet hose and the water pump. To reverse flush, force a garden hose into the end of the outlet hose and aim the inlet hose so water doesn't splash on the engine. Turn on the water; it should flow through the garden hose and come pouring out the inlet hose. When the water runs clean for two minutes, the job is done. Reconnect the hoses and tighten the clamps.

If the water just trickles out of the inlet hose, the heater core is badly clogged. The water pressure may (after a few minutes) clean the passages; if it does not, reconnect everything and take the car into a radiator shop for professional service.

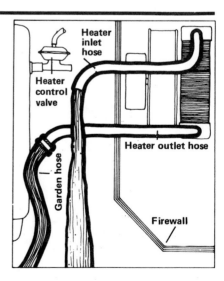

Heater inlet hose

Heater control valve

Garden hose

Heater outlet hose

Firewall

mended that you attempt to install one of these yourself, but when professionally installed they can be very effective. Some new American cars offer an in-block cooling-system heater as a factory option. If you're in the market for a new car, consider taking that option.

In addition to instant heater output, electric coolant heaters offer some fringe benefits. They make the car much easier to start in cold weather. Except for the battery, the car thinks its summertime, since all engine parts and the oil are at normal operating temperature. Also, they contribute to longer engine life, since much engine wear occurs during cold starts and before warmup.

Battery Maintenance

Cold weather really gives a battery a tough workout. The lower the outside temperature, the less power a battery can develop; in wintertime it is not uncommon for as much as 60 percent of battery power to be lost. The problem is compounded by the fact that at low temperature the engine oil is thicker, and the starting motor must spin the crankshaft through a pan full of thick oil. This takes a lot of extra cranking effort.

To give your battery the best chance to do its job, you should perform some simple maintenance tasks; you might also consider a battery warmer in very cold weather.

One of the most common battery-related causes of hard starting in winter is neglect of the cable terminal connections. A high percentage of so-called dead batteries are really adequate; the current just can't flow through the corroded connections fast enough to crank the engine properly. Keep the terminals clear of corrosion and you will have fought more than half the battle.

If a battery runs down, it is likely that something is wrong with the car's charging system. If the dashboard light or meter doesn't indicate trouble, and the battery cables are clean and tight, check the drive belts at the front of the engine. Press down on the belt that's

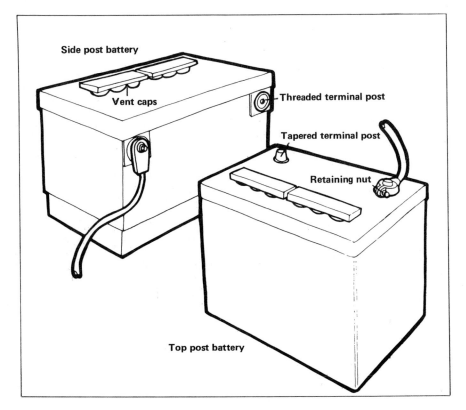

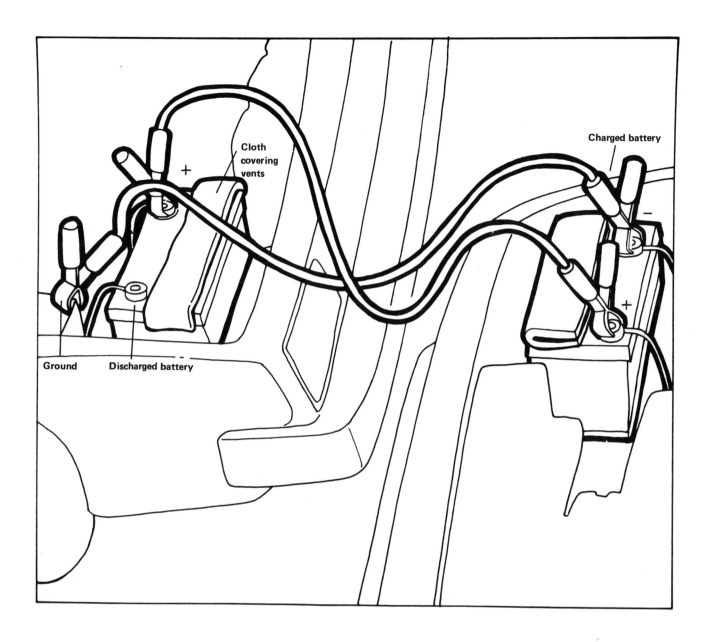

Charged battery

Cloth covering vents

+

−

−

+

Ground   **Discharged battery**

Working on Your Battery

Car batteries are basically simple devices, but if you are working on your battery, either giving it preventive maintenance or trying to connect a trickle charger or a set of booster cables, one problem you might encounter is determining which is the negative terminal and which the positive. Do not begin any sort of work until you are sure you know which is which.

On many batteries, there's a plus sign (for positive) and a minus sign (for negative) embossed on the battery case right next to the terminal. If there isn't, here are two other ways to tell:

With a top-post battery, look at the battery posts. The one with the larger diameter is the positive one. The difference in diameter is rather slight, but noticeable.

Follow the cables from the battery posts. One (it may split into two cables) is simply bolted somewhere on the body and/or engine. The other cable goes to the starter motor, which is on one side of the engine at the bottom rear. On most Ford and some American Motors cars, the other cable is only a few inches long and goes to an electrical component—a relay—mounted on the sheet metal of the engine compartment. If you do not know a starter motor or relay from a crankshaft, don't worry. The cable is connected to the starter or relay by a nut, while the other cable is bolted in place. The first connection will be an obvious electrical one; the second will look as if the cable is merely being bolted to hold it in place. If you can see those differences, you can tell which is which. The cable to the starter or relay is the positive one; the cable bolted to the body and/or engine is the negative one.

wrapped around the alternator (the belt-driven component with the wiring connected to it). If the belt deflects less than 1.9 cm (3/4") under thumb pressure, midway between the alternator pulley and the nearest other pulley, it's okay. If it deflects too much, it should be inspected by a mechanic. Perhaps it just needs retightening.

In many cases, the battery weakness is caused by frequent short-trip demand, such as a couple of miles to a bus or train station in the morning and the same at night. The alternator doesn't get a chance to fully recharge the battery. If the weather is really cold, the battery not only doesn't operate at full capacity, but it takes extra energy to get it started (unless, of course, you have installed electric heaters). The battery warmer does not take the place of a coolant heater; you could use both to advantage in very cold weather. If you don't want the complex and expensive coolant heater, the battery warmer might be enough to help you with winter starting problems. It ranges from $2 to $18, and is easy to install yourself.

There are two types of battery warmers. One is a plate type that fits in the battery box. Just undo the bracket that holds the battery, disconnect the cables and lift the battery. Clean out any debris on the base of the battery box, place the plate in position and refit the battery. Plug into household current overnight. The typical plate warmer draws 50-60 watts.

The other is a miniature electric blanket that wraps around the sides of the battery. The typical design has a vinyl cover filled with thick fiber glass insulation, so it helps even if you can't plug it in, as when the car is left in a parking lot, at work, or a train or bus station. The blanket type comes in sizes to fit different batteries; wattage ranges from 80 up.

Another possible approach to the problem is to attach a small battery charger, called a "trickle charger," to the battery overnight.

Trickle chargers, so-called because they deliver a trickle of electricity to the battery, could take days to fully charge a dead

---

Doing It Yourself: Battery Maintenance

You Will Need:
wrench
wire brush
petroleum jelly
baking soda
cable terminal (nickel-plated type); hacksaw; cable-terminal puller (optional)

There are two types of battery cable connections—the top post and the side terminal—and both require maintenance. Special wire brushes, designed specifically for one type or the other, are available in most auto-supply stores for about $1 and up.

Doing the Work

Top Post Batteries: The most common battery, with cables that connect to cylindrical posts on the top cover. White corrosion deposits are easy to see.

To remove the cable terminal for cleaning, slacken the retaining nut, then manually twist the terminal back and forth to free it up. Continue to twist while pulling upward, and it will come off.

If the terminal is stuck, tap it with a hammer, first on one side, then on the other, until it moves. If it still won't budge, obtain a cable-terminal puller, an inexpensive device that has jaws to grab the underside of the terminal and a forcing screw that is turned down against the battery

post. As you turn down the screw, the jaws lift up the cable terminal. This puller is available in most auto-supply stores, for about $2.

Once the cable is off, go over the post and both inside and outside of the terminal with the wire brush. Then refit the terminal to the post, gently tapping it down with a hammer if necessary. Tighten the terminal nut and smear the entire connection with petroleum jelly, which will help retard corrosion.

If the cable terminal is so badly corroded that it virtually disintegrates as you remove it, forget about cleaning. Obtain a replacement cable terminal (about $.50) from an auto-supply store. Remove the old terminal with a hacksaw, strip away 1.3 cm (1/2") of insulation from the cable end and slip the exposed wire into the new terminal. Tighten the nuts and bolts on the replacement terminal, then install on the battery post, and coat with petroleum jelly.

If the top of the battery is very dirty, the battery won't hold a charge very well. Clean with a solution of baking soda and water (follow the proportions given on the baking soda box for general cleaning).

Side Terminal Batteries: Corrosion on the side-terminal battery cable connections is difficult to see because the mating faces that form the connection are not exposed. This type of terminal does have better resistance to corrosion, though it is far from immune. The contact area is much

smaller than the top post design, so a smaller amount of corrosion can have a greater effect.

To service, remove the bolt that holds the cable terminal to the battery. It's a small one and often quite tight, so you need the right size wrench—use a 7.9 mm (5/16-inch) box, open-end or socket rather than an adjustable wrench. Wirebrush the mating contact faces thoroughly, paying particular attention to the recesses in the notched areas.

When they're brushed clean, just refit the cable, install the screw, start the screw in by hand, then tighten securely with the wrench.

Adding Water: Although many new cars have sealed batteries, most batteries still require periodic addition of water. Distilled water is universally prescribed for battery use, but the plain truth is that although distilled water is nice, it's rarely available. If you have some in the house, perhaps for use with a clothes iron, that's fine. But don't be afraid to use plain tap water, which is what's in those service station fillers. Do not use expensive bottled drinking waters; in most cases they are mineral waters, and minerals are not good for batteries (hence the distilled-water recommendation). The odds are that your tap water has far less mineral content than bottled water.

battery. Used to keep the charge level up, however, they work very well. They are often on sale at discount stores for less than $10, and the smallest (two-amp rating) will be adequate in most cases.

The charger is easy to attach—if you know how. The wire with the black clip is always attached to the battery's negative terminal; the wire with the red or green clip is connected to the positive terminal. Turn the charger knob or switch to 12V (for 12 volts, the voltage of your car battery) and plug in the charger; that's all there is to it.

Tires

Your car's tires are your only contact with the road surface when you are driving. The condition of the tread and the degree of inflation of a car's tires are always of enormous importance, but perhaps never more so than in the winter, when the condition of the road surface is frequently less than ideal. Ice and snow do, of course, make an enormous difference in tire traction; stopping distances are decreased significantly. This means you have to drive with that knowledge well in mind and that you should, and at time will be required by law to, use special tires or attachments to the tires to suit the driving conditions.

There are a number of choices available to you for winter driving; unfortunately, none is perfect, which is why care must be coupled with whatever else meets the road.

Snow Tires

These are tires with treads specially designed to increase traction and handling ability when there is snow on the road. Snow tires can be put only on the driving wheels or, for somewhat better handling (particularly if you have front-wheel drive), on all four wheels. If you use only two snow tires, they must be of the same construction as the other tires; that is, radial tires must be used with radial snows, bias-ply with bias-ply snows. There is one exception to this rule: belted bias-ply and simple bias-ply are compatible, so it is possible to use bias-ply snow tires (which are less expensive) with belted bias-plys.

Studded Snow Tires

These are snow tires with carbide tips embedded between the treads to improve traction, particularly on ice. Although regular snow tires do not significantly shorten stopping distances when there is ice on the road surface, studded snow tires do. When they were first introduced, in fact, studded tires were hailed as the answer to winter driving problems. Unfortunately, it is now believed that studs inflict damage on the road surface, and a number of states have banned their use. At present studded tires are not permitted in California, Connecticut, Hawaii, Illinois, Louisiana, Michigan, Minnesota, Mississippi, Pennsylvania, Rhode Island, and Wisconsin. In twenty-seven other states, there are limitations on the months when studded tires can be used. Before you invest in studded snow tires, check the law in your state and any other state in which you may be traveling.

Radial Tires

There is a moderately widespread misconception that radial construction tires give you enough road grip and handling to replace snow tires. This is simply not true in any but the lightest snowfall and most minimal accumulation.

If winters in your area are quite mild, however, a quality radial tire will offer traction in *light* snow close to that of an ordinary snow tire. If your car has front-wheel drive, the superior traction of that system, particularly when combined with good radials, will meet most mild winter conditions.

All-Weather Tires

Several European tire manufacturers and a handful of American ones are coming out with a so-called all-weather tire, with the tread specially compounded to "grip" the road even on ice, snow, and slush. They are marketed under various names, and tend to be a good deal more expensive than snow tires.

Extra Wheels

If you buy snow tires, buy a pair of spare wheels for them. Snow tires last a long time, for most motorists do far less driving in winter than in other seasons. If you plan to keep your car for a few years, a set of spare wheels (you can frequently get them at a wrecking yard for $10 each) is a good investment. The cost of the wheels is recovered by the saving in price and/or time of mounting and demounting them each year.

The tires should last longer, too, because they will not be subject to mounting and demounting damage. When you remove the wheels to store the tires for the warm weather, lower air pressure to .7 - 1 kilogram per square centimeter (10 - 15 pounds per square inch) and lay them flat on a piece of board.

Note: If you have studded snow tires, they must be remounted each year in the same position. When you take them off, mark them with a piece of chalk (R for right side, L for left side).

Chains

Under certain conditions, chains are the only safe solution, whether you have snow tires or not. In deep snow or ice, chains provide significantly better traction than snow tires. They reduce stopping distances on ice and packed snow. Many people object to chains

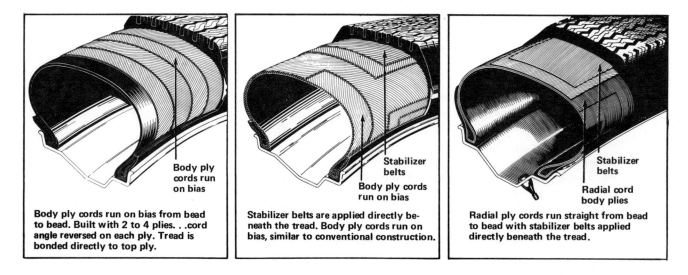

Body ply cords run on bias from bead to bead. Built with 2 to 4 plies. . .cord angle reversed on each ply. Tread is bonded directly to top ply.

Stabilizer belts are applied directly beneath the tread. Body ply cords run on bias, similar to conventional construction.

Radial ply cords run straight from bead to bead with stabilizer belts applied directly beneath the tread.

on numerous grounds, however. Unlike snow tires, which can be used on any road surface, chains must be removed when roads are clear and then replaced when road conditions warrant their use. Although some improvements have been made in the ease with which chains can be installed, it is at best a pain in the neck to put them on. Nor can you drive at "normal" speeds when you have chains on your tires. Forty to forty-eight kilometers per hour (twenty-five to thirty miles per hour) is the maximum recommended speed for most chains. Of course, if the driving conditions warrant the use of chains, you probably shouldn't be driving much faster than that anyway. There are some new types of chains made of polyurethane that the manufacturer claims can be driven at highway speeds.

Another objection to chains has been that you cannot use them with radial tires. No longer true. Chains especially designed for radials are now available.

If you balk at the idea of carrying a set of chains in your trunk during the winter months, consider a pair of clip-on chains. These are not a full tire chain; rather they clip on to a section of the tire, and are designed for emergency use, simply to get you out of a tight spot. They cost just a couple of dollars and are a sensible addition to any car emergency pack.

Before you purchase chains, by the way, check to be sure your car can handle them. This is not a widespread problem, but there are a few models of new cars with insufficient clearance in the wheel wells to accommodate chains. Owners of such cars have no choice but to use snow tires.

The Body

The increasing use of ice-melting salts on highways means that the road film that hits your car's body paint and underbody will promote rusting. Removing snow from the body with a stiff brush also adds to the damage. Garaging the car for the entire winter is not a practical answer for most people, so here are some alternatives:

Have you car body rustproofed. The rustproofing service, $100 to $150 for most cars, will help prolong its life. If you're handy, you can buy do-it-yourself rustproofing kits and do the job for about $60.

Clean and wax your car before winter hits its peak, and again during a thaw around midwinter.

At frequent intervals between periods when there is snow and slush on the road, take your car to a clean-your-own-car wash and use their hose to spray under the car, in and around the wheel wells, under bumpers, and other places where frozen, salty slush may have collected. You could do it with your own hose, but car-wash water pressure is usually stronger and you can do a more thorough job.

If you can, keep the car in a garage when not in use. If you can't, consider a car cover. Covers range in price and design, but their virtues may not outweigh their disadvantages. A cover can be inconvenient to remove, particularly if snow is piled on it. In addition, you must be sure the car is clean and dry before refitting the cover or you risk damaging the paint job. If you have a car you really love, however, and must park it outside, you may be willing to put up with the inconvenience.

If you have a convertible or sports car with a flexible plastic rear window, keep the top up all winter. Resist the temptation to put on a ski mask and do some top-down touring, for in cold weather the plastic will harden and probably crack.

If you have a motorcycle, scooter, or moped, wax the tubing

and sheet metal, and apply a film of chassis grease (available in tubes in auto supply stores) to pedal shafts and control cables.

Winter Vision

Maintenance of the heater/defroster is a major aspect of ensuring good winter vision, but when there's thick ice on the windshied, you can't expect the defroster to melt it instantly, even if the coolant is warm. Even if the defroster could do the job, it would take entirely too long to melt thick ice.

Deicing spray in collaboration with a plastic scraper is one popular answer. But because deicer is basically just antifreeze and a propellant, you should be careful to spray on the windshield only, keeping it off the body paint. And always use a plastic scraper, never a putty knife, screwdriver, or other metal object.

Heat guns that plug into the cigarette-lighter socket are another possibility, but they are somewhat slower than deicing spray and they draw battery current. If you get one of these guns be sure to start your engine first, so the charging system can go to work.

Some people have been observed trying to deice the windshield with pots of hot water. At best, it takes a lot of hot water to melt thick ice, so unless you keep the pots coming out in a steady stream, the hot water itself will probably freeze. At worst, you can end up with a shattered windshield. Another bad idea is using ice-melting salts designed for road use. Most of it will end up on the body paint and do it no good.

Another approach is to try to keep ice and snow off the windshield to begin with. Windshield covers—rubberized canvas sheets with magnets in the corners—will do a fair job, but frost can still accumulate under the cover. The amount of scraping required will be reduced but not altogether eliminated. Avoid the cheaper plastic covers, which may stick to the frost that is stuck to the windshield; you will end up with a bigger job than if you had no cover at all.

Factors That Affect Fuel Economy

Temperature
Summer temperatures (over 70°F) [21.1°C], are better for fuel economy than winter temperatures. At 20°F [6.7°C], for example, there can be an approximate 8 percent fuel-economy loss. For a 20-miles-per-gallon [7.1 kilometers per liter] (city and highway combined) vehicle, this is about 1.5 MPG [.53 KmPL].

Wind
Wind can increase or decrease fuel economy. Examples for a car that normally gets 20 MPG [7.1 KmPL] (combined) are:
18 MPG [6.4 KmPL] tailwind—about 12-percent gain in fuel economy (2.4 MPG) [.85 KmPL].
18 MPG [6.4 KmPL] crosswind—about 1-percent loss in fuel economy (0.2 MPG) [.07 KmPL].
18 MPG [6.4 KmPL] headwind—about 10-percent loss in fuel economy (2 MPG) [.7 KmPL].

Precipitation
Rain or snow, and the wet roads that result, can cause an approximate 10 percent loss in fuel economy (2 MPG for a 20 MPG vehicle) [.7 KmPL for 7.1 KmPL vehicle].

Road Condition
Rough or loose road surfaces (such as sand or gravel) can also cause a fuel-economy loss ranging between 10 and 30 percent (or 2 to 6 MPG for a 20 MPG vehicle) [.7-2.1 KmPL for 7.1 KmPL vehicle]. Cars use more fuel on hilly roads than flat roads. The fuel saved in going downhill does not equal the extra fuel used going uphill. Mountain driving causes an even greater fuel economy penalty.

How You Drive
An engine that is already warmed up (such as one that was used in the last four hours) requires less fuel to reach its most efficient operating condition than a "cold" engine (such as one in a car parked overnight). Trip length also affects fuel economy. Shorter trips (under 5 miles) [8 km] do not allow the engine to reach its best operating condition, whereas longer trips allow the peak operating temperature and engine condition to be obtained. This does not mean that you can save fuel by increasing the length of your short trips. It does mean that by combining numerous short trips into a single, longer trip you can save fuel by reducing the total miles driven as well as taking advantage of your vehicle's warmed-up condition. Smooth, even driving improves fuel economy performance; therefore, try to avoid sudden stops and starts. By anticipating stop lights and intersections, you can slow down gradually. Also, avoid rapid accelerations. On the highway, you will improve your fuel economy by driving at or below the 55-MPH [88.5 KmPH] speed limit.

Your Vehicle's Condition
The condition of your vehicle is important, too, for fuel economy reasons:
- Maintain your vehicle according to the manufacturer's specifications. On the average, a tuned-up vehicle gets approximately 3 to 9 percent better fuel economy than one that has not been properly maintained.
- Keep the tires inflated to the proper pressure. Underinflated tires can cause a fuel economy loss.

1977 Gas Mileage Guide — FEA/D—77/007
U.S. Government Printing Office:
1977-0-228-321

Windshield Wipers

You might get by with malfunctioning windshield wipers in a mild drizzle, but if they don't work in a driving snowstorm, you can be in real trouble.

If your car has the disappearing type of wiper, which rests in a recess just below the hoodline, some special attention is required. Snow can pack around the wipers, and when the temperature is very cold, the wipers could stick in place. This often happens when the car is parked. If you have recessed wipers, check them and dig out any ice before you attempt to drive away. Don't move the car until you have the wipers working. If the wipers stick, they may draw enough current to blow a fuse.

If your wipers streak the windshield, they may just be greasy, or they may be near the end of their useful life. Try cleaning the blades; some alcohol on a clean rag works fine. If that does not improve matters, you may have to replace the blades.

Wiper blades rubber inserts, which do the actual wiping, are aged quickly by summer heat; by the time winter comes, they may have hardened and developed tiny cracks in them. Check the entire wiper blade. If it isn't bent or cracked, you probably just need new rubber inserts, which are about a third the price of new blades. Installing them is a simple procedure.

If the wiper blade appears bent, get a new blade. To detach the old one, look over its connection to the wiper arm carefully. You'll see a locking tab or depressable button, which will enable you to take off the blade.

If winters in your area are severe, you may find that even new blades will not perform properly. The solution is a so-called arctic blade, which is encased in a special rubber that flexes smoothly in even the severest weather.

Washer System

The windshield washer is another important visibility aid in winter driving. If your reservoir is filled with a summer cleaning solution (perhaps simply water), empty it and refill with washer "antifreeze," a special solution with some alcohol in it to prevent it from freezing. Check the washers; if they malfunction, have them serviced before you need them. Incorrect aim of the nozzles, a common problem, is something you can correct yourself. On some cars the nozzle is threaded into place and need only be twisted by hand or with pliers to reset it. On others, the nozzle is a thin tube held in place with a clamp and screw, and is accessible from under the hood. Bend the nozzle tip as required to get it squirting where it should.

If there is only a tiny squirt from a nozzle, perhaps it is clogged. Take the car to a service station and ask to have some compressed air blown through the nozzle. Do not try to unplug it with a needle, which could snap off in the nozzle and make things considerably worse.

Rear-Window Defroster

If you bought your car with a rear-window defroster, you know what a good investment it is, particularly for winter driving. If you did not take that option and are thinking of buying a defroster kit to add on, there are two types of rear-window defrosters. One, which looks like the factory option, is a printed circuit grid that attaches to the glass and is wired to a dashboard switch and the car battery. The problem: It takes a lot of current to operate a circuit board that will do the job quickly. When you buy a rear-window defroster as a new car option, you get a high-output alternator and a heavy-duty battery as well. That's why the defroster option costs about $100, while the kits sell for as little as $7. If you merely install the kit defroster, you may run down your battery, particularly in short-trip operation. Of course, most kits don't draw nearly as much current as the factory option, but then they don't do anywhere nearly as quick a job of cleaning the rear window.

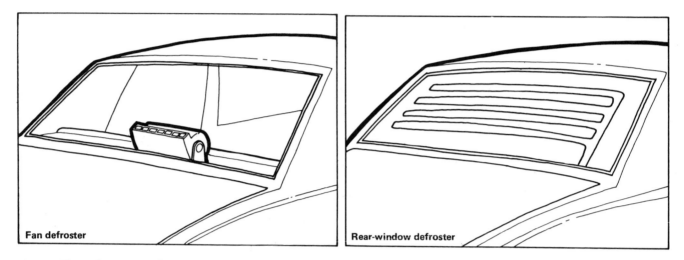

Fan defroster

Rear-window defroster

The other type of rear-window defroster is a fan that blows warm air from the passenger compartment onto the glass. It too is wired to a dashboard switch and the car battery. This type draws very little current, however, so it can be installed without your worrying about its effect on the battery. Like the circuit board kits, it works very slowly.

If you start out on a trip with a clean rear window—having used a scraper and perhaps some spray deicer—one of these kits will probably keep it clear, but it won't do any primary cleaning to speak of.

Lights

You have not completed your winter maintenance check if you have neglected your car lights. Check that the headlamps are properly aligned, both high and low beams, so that you light up the part of the road you need to see. Parking lights and turn signals must be in proper working order too. Of course, this is true for all seasons; but in winter, when visibility may be especially poor, your lights take on even greater importance.

Throughout the winter season you should frequently wipe off your car lights, front and rear. Spatters from wet and slushy roads can coat them with a film that will reduce their illuminating power.

Fuses

Most essential services in your car are controlled by fuses. When something isn't working—lights, windshield wipers, heater, etc.—the first thing you should suspect is a blown fuse. Inform yourself of the whereabouts of the fuse box in your car; it is most likely under the dashboard, but check your owner's manual to be sure. It's a good idea to keep a spare set of fuses handy, as well as a flashlight, since the fuse box is usually in a dark place (often made darker by the lost connection). If you like to be ultraprepared, some fall day when you have nothing else to do, you can inspect the fuse box and make yourself a wiring diagram so you'll know for sure which fuse corresponds to which function; it may save you some time should a vital fuse blow when the weather is less pleasant.

Exhaust System

In any season, your exhaust system should be in top working order, but in winter a faulty system can be particularly dangerous: windows are more frequently kept closed when it's cold out, and ventilation in the passenger compartment is limited.

If your car's exhaust system is noisy, you probably have a leak. To check, try to hold a rubber ball against the tailpipe opening while the engine is idling. If the pressure build-up makes this impossible, or if the engine stalls, the exhaust system is reasonably tight. If you can hold the ball there with no problem, you have a leak and should get it serviced.

In any event, it's good winter practice to always drive with the windows open at least .7 cm (1/4"). This reduces the danger from fumes if the exhaust system is leaking. Even a small leak can be dangerous.

Hitting the Road

Getting Into Your Car

Before you can start your car you have to get in it, and in winter that may be a significant problem. If it's safe to do so, leave one door unlocked. If you can't, keep a small can of aerosol lock de-icer handy, but not in your glove compartment, where you won't be able to get to it. If you don't have any deicer, do not use automobile antifreeze. Instead, heat the key with a cigarette lighter or match, insert, and try to move the lock. Don't force it or you may break the key. Just let the heat transfer from key to lock, remove the key and reheat. Eventually, the lock will defrost. As a last resort, unbend a coat hanger, shape an end like the letter U, and force it through the top of the window into the passenger compartment. Hook it around the door-lock knob and lift up, just as you would do if you'd lost your keys. It may take several minutes, but replacing a broken door lock or removing it because a key is broken off inside is a job for a locksmith, and the price will not be negligible.

Winter Starting Technique

Starting a car in winter is different from other times of the year. Everything is cold and sluggish, and unless you do it right, you may find your engine flooded or your car stalling. The correct procedure for most cars begins with pressing the gas pedal firmly to the floor, then releasing it completely. This will properly position the automatic choke and the fast-idle linkage. The choke is a plate that restricts air flow through the carburetor, making the gas-air mixture richer with fuel, which is helpful for starting. The fast-idle linkage automatically sets the carburetor so the engine will run fast when it is first started, thereby helping prevent stalling.

Turn the ignition key, and when the engine turns over and starts, let it run for half a minute before you proceed. If it stalls immediately, repeat the pedal-flooring procedure. If the engine does not catch, do not keep cranking it for more than fifteen seconds. Release the key, wait a minute, and then try again.

In very cold weather and/or if your car has been standing unused for several days or more, begin by pumping the accelerator pedal (pushing it to the floor and releasing it rapidly) two or three times. This will operate the accelerator pump, a device in the carburetor that adds a few extra squirts of gasoline to the mixture. Crank the engine; when it starts, let it run for half a minute before driving away.

Do not pump the gas pedal as a routine starting procedure in winter. Unless the weather is extremely cold, the extra gas may be too much and the engine may stall as a result. This is the condition commonly called flooding. The smell of gasoline will tip you off that the engine is flooded. In that event, wait a few minutes, then press the gas pedal firmly to the floor and keep it there. This will open the carburetor linkage at the choke and permit more air to enter. Crank the engine and it should start.

If your car has fuel injection, ignore the above instructions. Just turn the key; *do not* touch the accelerator pedal.

Do not get frazzled if your car does not start up right away. Starts do put a drain on the battery, and it is not a good idea to overtax it, especially in winter. Start the usual way to begin with; if the engine turns over but doesn't catch, continue trying as described above no more than four or five times. Wait a few minutes between attempts to give the battery a chance to regain some power. Do not simply slam down the gas pedal and keep turning the key; the damage you do will go beyond the battery.

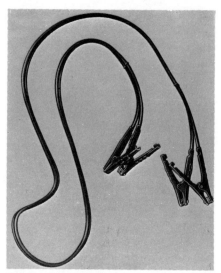

Battery charging cables

Doing It Yourself: Automatic Choke Check

If your engine cranks vigorously but won't start, make this simple check of the automatic choke:

Remove the cover from the air cleaner, the big sheet-metal cannister on top of the engine. On most cars the cover is held by a wing-nut that you can simply twist off by hand. On others it may be held by a couple of ordinary hex-head nuts or bolts, which can be removed with an adjustable wrench.

Inside the cannister is the air filter element, a cylinder of pleated paper in a metal or metal-and-plastic cage. In the center of the air-cleaner housing you'll see a large opening for the carburetor, the part it's mounted on. The typical carburetor has one, two, or four large vertical bores, called barrels; hence the terms you may have heard—one-barrel, two-barrel, or four-barrel carburetor. At the top of one or two barrels is a plate mounted on a pivot shaft, so the plate can pivot horizontally to cover the barrel, or vertically to permit maximum air flow. This is the choke plate, and when the engine is cold, the plate should be almost horizontal to close off the barrel. If the plate is not covering the barrel completely, the engine will be hard or impossible to start in cold weather.

If you encounter a vertical or nearly vertical choke plate after flooring the gas pedal for a cold start, something is sticking. Try pivoting the plate closed with your fingers, and if it springs back open, hold it in place with a long rod while a helper tries to start the engine. The object of using a rod is to keep your fingers away in case the engine backfires through the carburetor.

If the engine starts with the choke plate pushed closed, take the car to a mechanic and ask him to free up the choke. It should be an inexpensive job.

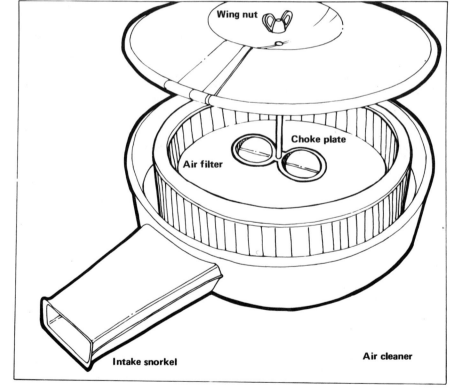

Wing nut

Choke plate

Air filter

Intake snorkel

Air cleaner

Starting Fluid

When the temperatures are colder than normal, the carburetor will have trouble vaporizing the fuel into a combustible spray. Gasoline is formulated for different geographical areas; the gasoline that would start a car in Florida would not necessarily start one in Maine. If the weather is abnormally cold, however, the Florida car might well need a gasoline formulated for a colder climate.

If you encounter hard starting when it's unusually cold and the engine cranks normally, perhaps fuel vaporization is the problem. If the choke is closing properly, the air supply is reduced and what little fuel does vaporize will mix with what little air can enter and the car should start. *Should* does not mean *will,* and *start* does not mean *run.* The engine may not start, and even if it does, it may instantly stall out.

Before you run down the battery, you may wish to try a spray starting fluid, a product that is primarily ether, which will vaporize even at $-54^\circ C$ ($-65^\circ F$). This fact makes it very flammable, so handle with care.

Remove the air-cleaner cover, push open the choke plate and spray into the carburetor for a few seconds (up to three). Let the choke plate close, then try to start the engine. *Do not* spray while a helper cranks the engine; it could backfire and give you a nasty burn.

If the engine turns over but very sluggishly, there are two likely sources of trouble: the battery and heavy oil in the engine.

The battery problem may be caused by corroded cable terminals, which you can service yourself (see page 153), or the battery may be weak, in which case a recharge or boost-start (see below) will solve the problem. Even if you have thick oil in your engine, a good boost may get you started and permit you to get an oil change at your convenience.

Once the engine starts and has been running for half a minute, drive away at moderate speed, less than 64 KmPH (40 MPH). Do not

How To Start a Car with Booster Cables

Booster cables are a handy item to have for starter emergencies; if your car battery is more than two years old, they are an essential item. Available in supermarkets, discount stores, and auto-supply outlets, a set of two cables costs between $3 and $10. The best type is six-gauge, all-copper wired.

The only other thing you need is another car with a healthy battery of the same voltage as the dead one; the cables do not themselves recharge the battery, they simply carry the charge from the strong battery to the weak or dead one. Jump-starting a car is basically a simple and safe procedure as long as certain safety precautions are taken.

1. Do not let any part of the two cars touch; take special care that the front bumpers are not in contact.
2. Before you do anything else, turn off the ignition and all accessories in both cars (radio, heater, lights, windshield wipers, etc.). Engage parking brakes and put both cars into neutral or park.
3. Remove the cell caps from the dead battery and check that the battery fluid is not frozen. Do not attempt to jump-start a frozen battery; it could explode. If the battery is frozen, remove it to a warm place to thaw out gradually. If the fluid is still in a liquid state, cover the filler holes with a clean rag. If fluid level is low, refill with water before proceeding.
4. Connect positive terminal to positive terminal via the cables, but do not make the final negative-to-negative connection. Rather, clip one end of the cable to the negative terminal of the working battery and clip the other end to an engine bolt at least a foot away from the battery. Take utmost care not to make a positive-to-negative connection.
5. At no point should you permit the cable clamps to touch each other.

Once you have made your connections, start the booster car and keep it running moderately fast. You need not race the engine, but do not let it stall. Boost-starting draws a tremendous amount of current from the good battery, so you want the engine running fast enough to start recharging immediately.

Crank the engine of the car with the dead battery. As soon as the car starts, unclip the booster cable from the engine bolt, then unclip the other end from the negative terminal. Next, disconnect the booster from the positive terminal of the boosted car, and finally from the positive terminal of the booster car. Be careful at all times to hold the spring clip away from all metal parts of both cars.

attempt to let the engine warm up by idling, for this causes unnecessary wear. If you simply drive away, the engine is placed under a mild load, which speeds warmup. The faster the engine warms up (within limits), the better it will perform. Parts reach operating temperature more quickly, and when they are at operating temperature they fit and work together better, with less friction.

Full Gas Tank

Before you pull into your driveway for the night, check the fuel level. If the gauge reads less than half, fill 'er up. The closer the gas tank is to full, the less chance there is that condensation in the gas tank will form ice particles that could make the car difficult or impossible to start, and/or cause repeated cold stalling when you drive away.

An alternate possibility is to pour a can of fuel-system antifreeze into the tank. This special alcohol compound keeps the condensation from freezing. It's not expensive, but putting gas in the tank—something you'll have to do anyway, eventually—is even cheaper.

Winter Driving

Even if your car has been well maintained and you have managed to get it started without running down your battery, winter means cold, precipitation, and hazardous road surface conditions to contend with. A lot of care and a bit of knowledge about how to handle a car in winter will go a long way toward making the going safer.

Traction, Stopping Distances, Skidding

When roads are icy or covered with hard-packed snow, your traction is reduced and stopping distances increase dramatically. Do not get yourself in a position where you have to stop quickly and surely; chances are you'll fail, and a skid or collision is the best you can hope for. Drive slowly, keep your distance from the car ahead of

---

Road Emergency Checklist

A sure way to minimize the effect of a road emergency in winter is to be prepared for it. Almost anything can happen to your car in winter, including getting stuck. If you have equipped it with studded snow tires or chains, the car's chances of making it through the winter without losing traction are improved; but there are no guarantees. And there are always such problems as a flat tire, which can be twice as much trouble in bitter cold and/or during a storm, on snow-covered or icy roads.

You can prepare yourself for many winter emergency situations by carrying as much of the following as possible in the trunk of your car.

- Flares—A package of three flares is the minimum, so you can place one every 45 to 60 meters (150 to 200 feet) behind your car. If you don't have flares, you cannot safely do anything on or around your car.
- A box of safety matches—To light the flares.
- A snow shovel—Digging away the snow may be the only way you can clear a getaway path. Don't settle for a garden spade, or you could spend hours shoveling and getting nowhere. A lightweight aluminum snow shovel is best.

- A 4.5 kilogram (10-pound) bag of sand—For traction if you're stuck in ice or snow.
- An operable jack, a lug wrench, and a flat board about 60 cm by 60 cm (2' by 2')—The flat board is a handy surface onto which to place the base of the jack, particularly if you are on a soft shoulder, into which the jack base would sink.
- A blanket—Especially if you travel with small children. When the car is stopped, the heater is off and a blanket will be more than useful. A sleeping bag, if you can spare one for the car, would be even better.
- Windshield scraper—The kind with a plastic scraper blade on one end and a stiff brush on the other is the safest and most effective. Do not try to scrape an ice- or snow-covered windshield with a makeshift metal blade of any sort.
- Deicer spray—An aerosol or pump container of an antifreezelike mix to melt windshield ice when it's too cold or there is not enough time to depend on your windshield defroster.
- Ether-based starting spray—For emergency starts (see page 163 and use with great care).
- A set of booster cables—For jump-starting your car (see page 163).
- Chains—If you have radial tires, be sure to buy radial chains.

- Tow rope or chain—In case your car needs to be pulled, or you need to pull another car, out of a tight spot.
- A pair of heavy work gloves—Remember, it will be cold outside, so a heavy-duty pair of gloves can help make the work a little less painful.
- A tarpaulin or large plastic garbage bags—Useful as wind protection and as a waterproof surface if you are doing work or walking outside in a storm.
- A coffee can—As a container to melt snow for emergency water.
  Inside the car, in the glove compartment you should carry:
- A working flashlight.
- A supply of extra batteries.
- A set of fuses to fit your car.
- A compass.
- Road maps for the area in which you are traveling.
- First aid kit—check it from time to time and replenish dwindling supplies.
- Some high-calorie, nonperishable food—perhaps some candy or dried fruit for emergency use only.

Special Techniques for Safe Driving on Ice and Snow

The National Highway Traffic Safety Administration has this to say about winter driving:

The best safety rule for driving on snow or ice is DON'T. Stay off the roads until they are clear—unless you absolutely have to drive. If you must drive, be familiar with the special techniques necessary to minimize the dangers involved.

Glare: Snow produces a glare which can adversely affect vision. The sun, shining on the snow, makes the problem worse. Keep a pair of sunglasses or yellow lenses in your car and use them.

Safe Following Distance and Fog: When driving under cold weather conditions, when roads are slippery, follow cars in front at a safe following distance. Increase your following distance to allow enough room to stop, if you have to. Also remember, with moisture on the ground (in the form of snow) you are apt to run into foggy conditions frequently. Fog, coupled with slippery conditions, requires more alertness, attention to maintaining a safe following distance, and driving with lights on low beam to improve your visibility and chances of your being seen by other drivers.

Braking: Know how and when to brake. When possible, use the braking power of the engine by down-shifting to a lower gear rather than by using the brakes. When you must brake, do not jam on the brakes—tap and release them in a pumping motion. Don't brake in the middle of a curve. If your vehicle goes into a skid, take your foot off the brake.

Skids: Don't panic. Don't oversteer. Don't jam on the brakes. Remove your foot from the accelerator. Turn the steering wheel in the direction of the skid; e.g. if the rear end of the vehicle is skidding toward the right, turn the steering wheel to the right. When you are able to regain steering control you may be able to resume braking by pumping the brakes lightly.

Traction: To retain traction and avoid skids—start out slowly if parked on a slippery surface. If your wheels start spinning, let up on the accelerator until traction is returned.

Before going up a hill, increase speed (within reason) to build up momentum to help you climb.

Before going down a hill, especially a steep one, slow down by shifting into a lower gear. Don't use your brakes going down a slippery hill.

When approaching a hill, either ascending or descending, observe other vehicles on the hill and how they are reacting to conditions. Stay well behind the vehicle in front so that you can go around it if it becomes stuck. If other cars begin to slide, spin out, or have to back down the hill, wait until you have enough room to maneuver before going up the hill yourself.

Use your *judgment.* By observing what other vehicles are doing, it may be apparent that the hill is too slippery and dangerous. Pull over to the side as far as you can without the risk of getting stuck, and wait for a salt or a sand truck.

Stuck in a Rut: The action you take depends on how badly you are stuck. Whatever you do, *avoid spinning your wheels,* since this will aggravate the problem.

If the snow is deep, shovel the snow from in front and back of the wheels (both front and back wheels). Also shovel out as much snow from under the car as you can.

If you have it, spread some salt or sand in front and in back of your driving wheels (or use traction mats if you have them).

Don't let anyone stand directly behind the rear wheels. While a little pushing assistance from a friendly passerby will often add that little extra momentum needed to get going, don't let anyone stand directly behind the wheels.

If you're using devices under the wheels for traction, or if the wheels dig into dirt or gravel, individuals behind the car may be injured by rocks or objects thrown rearward by the spinning wheels.

If possible, try to keep the front wheels pointed straight ahead until the car is moving. The rolling resistance of the front wheels is lessened when they are not trying to move sideways.

If your wheels keep spinning and the vehicle doesn't move, stop and let the tires cool. Tires heated from spinning will just dig you into a deeper rut.

If nothing works, try to rock the vehicle out of the rut by alternately shifting from reverse to second gear (cars with manual transmissions) or reverse to drive (cars with automatic transmissions). In cars with automatic transmission, check your owner's manual to make sure such a procedure can be followed with your particular car.

Some Additional Hints and Precautions:
Keep your tires inflated properly.

Make sure all passengers use their safety belts, including shoulder belts.

On slick roads, stay well behind traffic in front. Observe traffic coming toward you, and be prepared to take defensive action in the event oncoming cars go into a skid.

When your wheels begin to spin, let up on the accelerator. Often this will permit traction to return sufficiently to enable you to negotiate a slick spot.

Always keep your gas tank at least half full, and periodically add some "dry gas" to the tank. This will help to avoid gas line freezeup, and will help to rid your gasoline tank of moisture which could eventually build up and cause your gas tank to rust and leak.

In cold weather, especially if there is moisture in the air or if it has rained or snowed and then turned very cold, your parking brake can freeze. If this happens, try rocking the car (as described previously). Under such conditions it may be wise, when parking the car, to avoid setting the parking brake, relying on the PARK gear position (or reverse in cars with manual transmission) to keep the car from moving.

If your door locks freeze, try warming the key with a match or lighter before insertion into the lock.

Cold weather starting can be a problem with some cars. To minimize the possibility of your car's not starting, follow the manufacturer's recommendations for the oil to be used in winter (either a low viscosity or a multiviscosity oil). Follow the directions in your owner's manual for cold weather starting and remember, manufacturers recommend that cars equipped with catalytic converters should never be left to idle for extended periods to warm up, as this may cause the converter to overheat.

Safe Driving in Winter — U.S. Department of Transportation—HS 802 076 November 1976 you, and prevent skids by avoiding sharp turns or sudden braking.

If you do begin to skid, do *not* jam on the brakes. Remove your foot from the gas and steer in the direction of the skid: if your rear wheels are going to the right, turn the steering to the right. When the car begins to respond and you gain control of steering again, pump the brakes *lightly* until you have slowed down to a safe speed.

Winterizing Recreational Vehicles

Preparing a motor home or van for winter use is basically the same as winterizing a passenger car. If you plan to sleep in it in cold areas, such as ski country, check the weatherstripping around doors and windows. An air leak during the summer might not bother you, but trying to sleep when the cold and windy outside is infringing on the inside is the hard way to do things. Tubes of silicone rubber for repairing tears, or silicone cement for refitting loose weatherstripping, are inexpensive and simple to use (see Chapter 1).

If you are retiring a recreational vehicle for the winter, be sure to drain the waste and water tanks. Next, remove the spark plugs,

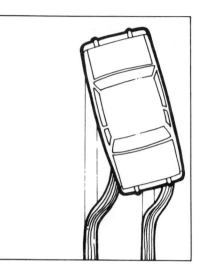

Turn wheels in direction of skid

Winter recreational vehicles

Road Emergencies: What To Do if You Are Snowbound in a Car

1. Stay in your car. It is your surest shelter; if you leave your car you may become disoriented and lose your way in a blizzard.
2. Make sure there is fresh air coming into the car. Remember that both snow and ice can quickly seal windows and doors shut, so leave a window open wide enough to give adequate ventilation. It is best to open a downwind window.
3. For warmth, keep the motor and heater going for about ten minutes every hour.
4. Make sure that your tailpipe is not blocked by snow; if it is, exhaust fumes will enter the passenger compartment. You will in all likelihood be unaware that this is happening.
5. The usual white distress flag will be of

little use in a snowstorm. Tie a brightly colored scarf or article of clothing to the antenna to alert rescuers. At night, turning on the dome light with some regularity will also be a signal for rescuers. If you have a flare, light it.
6. Exercise by moving your hands and feet. Do not remain in the same position for a long period of time.
7. If there is more than one person in the car, try to sleep in shifts.

—Ellen Leventhal

Road Emergencies: What To Do if You Must Leave the Car

1. Make sure that you are adequately covered. Wrap your feet in plastic, or any extra material or clothing you may have in the car. Cover your body also, using a wind-resistant material (such as plastic) if available.
2. Any wide, flat, light material can be fashioned into emergency snowshoes— thick evergreen boughs, a flat board, green boughs bent into the traditional shape for a frame, with rope, wire, or willow boughs for webbing.
3. Leave the car *only* if you know where you are going, and remember that walking in snow is exhausting—do not overexert yourself.

—Ellen Leventhal

squirt a few tablespoons of engine oil into each cylinder, then reinstall the plugs. The oil will coat the cylinders and minimize rust.

Remove the battery and place it on trickle charge (a trickle charger is $10 to $20 at auto-parts stores). After forty-eight hours it should be fully charged. Keep it in a cold place; when spring comes, it still will be fully charged. Note: Don't store a battery that isn't fully charged in a cold place, or it may freeze and crack.

Jack up all four wheels and support the vehicle by the front suspension and rear axle, using concrete blocks. This will take the load off the tires and prevent them from distorting.

If possible, run the engine out of gas. This will clear out·the fuel system and prevent gums that could make starting next spring impossible from foaming in the lines.

Small Gas Engines

If you have some small gas-engine appliances, such as an outboard motor, mower, chainsaw, lawn vacuum, or thatcher, they should be put away for the winter in a similar manner. This means squirting oil in the one cylinder, running the engine out of gas, and, if it has a storage battery, removing and charging it.

In addition, clean the body and remove debris from the underbody, using a putty knife. Clean around the wheel axles and coat the axles with chassis grease. Spray or squirt some penetrating oil on the throttle cable and linkage, find an indoor storage place, and cover with a piece of canvas.

Recent studies indicate that the average American is spending unnecessarily large amounts of money on the repair of the family car. This does not mean that we should all forget about servicing our cars and let them run themselves into the ground. What it does mean is that reasonable maintenance should be just that: reasonable. Much of what a car needs, winter and summer, can be done by the owner in a weekend afternoon. Most of the rest of keeping a car running well

---

Road Emergencies: What To Do if You Get Lost in the Snow

1. When travelling in snow-covered country, make note of landmarks and distance and time elapsed. These can be used to retrace your steps should you get lost. Remember that footprints disappear quickly if it's snowing.
2. If you are travelling in a group (and you should be under extreme weather conditions), stay with the group.
3. If you get lost, stop. Collect your thoughts, calm down, check maps, compass, and any other travel aids you may have. If possible, retrace your steps.
4. Because snow travel is difficult, it is better to stay put and signal for help. Any sound repeated three times should be understood as a distress signal.
   a. Blow on a whistle, honk your car horn, bang on metal with a stick.
   b. Build a fire. Adding rubber or oil will make the smoke black; water or damp or green wood will make a fire smoky and visible. A flaming fire is the best nighttime signal.
   c. Tramp out letters (SOS) or words (HELP) in the snow. Make them outrageously large. Fill the letters with leaves, twigs, moss—anything that will make them stand out in white sur-

roundings. If no suitable material is available, build snow mounds around the edges of the letters so that a shadow will be cast.
   d. Use a mirror or piece of metal to reflect the sunlight. At night, use a flare or flashlight. Signal SOS in Morse code.
   e. Lay out brightly colored pieces of cloth or clothing. If you hear a plane approaching, wave the cloth. If there is a tall tree and you can climb it, tie a brightly colored flag to its branches.
5. If it seems that there will be a long wait for help, build a shelter. Remember that snow is a pretty good insulator.
   a. Dig a trench, line the bottom with grass and twigs, cover the top with more branches, or, if you're lucky enough to have one, a tarp.
   b. Dig a cave into a snow drift. Line with branches, grass, etc.
   c. Snow or ice blocks can be used as an effective windbreak.
   d. Make sure that there is adequate ventilation in any shelter.
6. If no one knows where you are and there is little likelihood of a rescue party, you may have to travel. **Travel with caution.**
   a. Pick the easiest route, even though it may seem longer; it is important to conserve your energy. Go around obstacles, not over them; climb slopes in zigzag fashion, not straight up.

   b. Make emergency snow shoes
   c. Rivers or streams make good highways; but watch out for thin ice.
   d. Head for a known destination—look for signals of civilization, like smokestacks, highways lights, etc.
   e. Do not exhaust yourself, rest when necessary.

---

when it's cold outside is common sense. If you keep your car in good condition, have it winterized, equip it for emergencies, and drive with care, you and your car ought to survive the winter with no trouble at all.

Contributors

John Philip Baumgardt, former editor of *American Horticulturist* and author of four popular and successful gardening guides, is one of America's most distinguished gardening experts. He received a Ph.D. in botany with honors from the University of Missouri and is a member of three scientific honor societies. He lives on 500 acres of woodlands in the Ozarks, where he maintains vegetable gardens, an orchard, a bramble-fruit garden, a wild garden, annual flower beds, cold frames, and several acres of lawn. Mr. Baumgardt is on the editorial staffs of *Grounds Maintenance* and *Flower & Garden* magazines and is the garden columnist for the Kansas City *Star.*

Martina D'Alton, a freelance writer and editor, lives in New York City where she manages to keep warm each winter by running eight miles a day. She is the author of *The Runner's Guide to the U.S.A.* and has written articles on running for magazines such as *Travel and Leisure.*

Stuart Fischer has written for *The New York Times, Moneysworth,* and *The East Side Express* and contributed to *The First Complete Food Catalogue* and *Living Together in Tomorrow's World.* He now works for CBS News.

Nikki Goldbeck obtained her degree in food and nutrition from Cornell University. She is the author of *As You Eat, So Your Baby Grows/A Guide to Nutrition in Pregnancy,* a booklet published by Ceres Press, an independent publishing company started with her husband, David. The Goldbecks are co-authors of *The Supermarket Handbook* (Signet and Plume/NAL), a guide to buying whole foods; *The Good Breakfast Book* (Links/Quick Fox), a breakfast cookbook stressing high protein meals; and *The Dieter's Companion* (Signet/NAL), a guide to nutrition during all stages of life. The Goldbecks conduct workshops in nutrition for school and community groups including La Leche League, The International Childbirth Education Association, and natural childbirth classes.

Emily Jane Goodman is a New York attorney who frequently writes on the law. She is the author of *Tenant Survival Book* and co-author of *Women, Money and Power.* Ms. Goodman is now specializing in matrimonial, literary and entertainment law.

Barbara Kelman has edited numerous books, many of them on practical subjects. Born in New York City, she has lived in New England and currently resides on the windswept West Side of Manhattan. She finds keeping warm when it's cold infinitely more rewarding than keeping cool when it's hot.

Peter D. Lawrence is a freelance writer who lives and works in New York City. His articles on subjects as varied as world population growth, gemstones, and Asian food and customs have appeared in the *New York Times, Cosmopolitan, New York, Signature* and other magazines.

Ellen Schachter Leventhal, a freelance writer and researcher, has written on a variety of subjects. She lives with her family in New York City.

Jeanette Mall is a native of Colorado who now works in New York as an editor.

Bo Niles has written about architecture, building, and interior design for the last ten years, for a variety of shelter publications including *The Architectural Forum, House & Garden,* and *House Beautiful*'s Special Publications. Most recently, she was Architecture/Building Editor at *American Home* and a contributing decorating editor at *Sphere.* She is married with one son, and lives in New York City. When it's cold, she wears socks to bed.

Gary Olson is a researcher and writer who works extensively on a wide-range of subjects. He has a car that starts every single morning.

Alan Ravage is a writer, editor, and illustrator. He is currently at work writing and illustrating a book about cooking. A native New Yorker, he has had many years of experience keeping warm in such farflung corners of the world as Chicago, Illinois, and Milan, Italy.

Susan Restino is a writer who lives on a farm in Eastern Canada. She grew up in rural Connecticut, in a home full of gourmet traditions, and has since lived and gone to school in New York, Boston, London, Israel, and Vermont, before settling in Cape Breton, Nova Scotia, with her husband and two children. They raise goats, gardens, chickens, keep horses, work in the woods, do a lot of reading, and school the children at home. Her first published work, *Mrs. Restino's Country Kitchen,* (Link Books, 1976) is a blend of recipes, research on foods, and humorous advice on the ins and outs of country cooking.

David Sachs is a freelance editor and writer currently based in New York.

Douglas Smith is from rainy Seattle, and is naturally adept at combating damp and chilly climates.

Paul Weissler is a member of the Society of Automotive Engineers and International Motor Press Association. He has been writing about automotive technical subjects for 20 years. His articles have appeared in many magazines, including *Popular Mechanics* and *Mechanix Illustrated.* He is married and is the father of one child.

Suggested Reading List

Home Maintenance and Repair

Better Homes & Gardens Editors. **The Better Homes & Gardens Handyman Book.** New York: Bantam, 1974.

Castellano, Carmine C. & Seitz, Clifford P. **You Fix It: Insulation.** New York: Arco, 1975.

Cobb, Betsy and Cobb, Hubbard. **Vacation Houses: What You Should Know Before You Buy or Build.** New York: Dial Press, 1973.

Cobb, Hubbard H. **Money Saving Home Maintenance: The Complete Guide to Care and Repair.** New York: Macmillan, 1970.

Curry, Barbara. **Okay, I'll Do It Myself: A Handy Woman's Primer.** New York: Random, 1971.

Daniels, George. **The Unhandy Handyman's Book.** New York: Harper & Row, 1974.

Gladstone, Bernard. **The New York Times Complete Manual of Home Repair.** New York: Macmillian, 1966.

Hand, Jackson. **Complete Book of Home Repairs and Maintenance.** New York: Harper & Row, 1972.

Ingham, Andy. **Self Help House Repairs Manual.** New York: Penguin, 1975.

Popular Mechanics Complete Manual of Home Repair and Improvements. New York: Hearst Books, 1972.

Reader's Digest Complete Do-It-Yourself Manual. New York: Norton, 1973.

Sara, Dorothy. **Home Improvement and Maintenance.** New York: Macmillian, 1963.

Energy and Heating

Adams, Anthony. **Your Energy-Efficient House: Building and Remodeling Ideas.** Charlotte, VT: Garden Way Publishing, 1975.

Anderson, Bruce, with Michael Riordan. **The Solar Home Book.** Church Hill, NH: Cheshire Books, 1976.

Barber, Everett M. and Watson, Donald. **Criteria for the Preliminary Design of Solar-Heated Buildings: A Manual.** Guilford, CT: Sunworks Inc., 1975.

Clegg, Peter. **New Low Cost Sources of Energy for the Home.** Charlotte, VT: Garden Way Publishing, 1975.

Daniels, Farrington. **Direct Use of the Sun's Energy.** New York: Ballantine Books, 1974.

Daniels, George. **Home Guide to Plumbing, Heating, and Air Conditioning.** New York: Harper & Row, 1967.

Day, Richard. **The Home Owner Handbook of Plumbing and Heating.** New York: Crown, 1974.

Eccli, Eugene, ed. **Low-Cost Energy-Efficient Shelter for the Owner and Builder.** Emmaus, PA: Rodale Press, 1976.

Eisinger, Larry, ed. **Handy Man's Plumbing and Heating Guide.** New York: Arco, 1952.

Gay, Larry. **The Complete Book of Heating with Wood.** Charlotte, VT: Garden Way Publishing, 1976.

Halacy, D.S., Jr. **The Coming Age of Solar Energy.** New York: Harper & Row, 1963.

In the Bank. . .Or Up the Chimney? A Dollars and Cents Guide to Energy-Saving Home Improvements. Prepared by Abt Associates for H.U.D. Available from Superintendent of Documents, U.S. Government Printing Office, Washington, D.C.

Keys, John. **Harnessing the Sun to Heat Your House.** Dobbs Ferry, NY: Morgan & Morgan, 1974.

Kramer, Jack. **The Sun-Heated Indoor-Outdoor Room.** New York: Scribner, 1975.

Lucas, Ted. **How to Build a Solar Heater.** Pasadena, CA: Ward Ritchie Press, 1975.

Mother Earth News. **Handbook of Homemade Power.** New York: Bantam Books, 1974.

Murphy, John A. **The Homeowner's Energy Guide: How to Beat the Heating Game.** New York: Thomas Y. Crowell Co., 1977.

Save Energy Save Dollars. Efficient energy management for consumers from the Cornell Cooperative Extension Energy Task Force. Ithaca, NY: Cornell University, 1977.

Skurka, Norma and Naar, Jon. **Design for a Limited Planet.** New York: Ballantine Books, 1976.

Spies, Henry R., et al. **Three-Hundred-Fifty Ways to Save Energy (and Money) in Your Home and Car.** New York: Crown, 1974.

Springer, George S. and Smith, Gene E. **The Energy-Saving Guidebook.** Westport, CT: Technomic Publishing Co., 1975.

House Plants

Baines, Jocelyn and Key, Katherine. **ABC of Indoor Plants.** New York: Knopf, 1973.

Baumgardt, John. **The Practical Vegetable Gardener.** New York: Quick Fox, 1978.

Better Homes & Gardens House Plants. 2nd ed. Des Moines, Iowa: BH&G, 1971.

Cruso, Thalassa. **To Everything There Is a Season.** New York: Alfred A. Knopf, 1973.

Dworkin, Stanley and Dworkin, Florence. **The Apartment Gardener.** New York: New American Library, 1974.

Everett, T.H. **How to Grow Beautiful House Plants.** rev. ed. New York: Fawcett World, 1975.

Faust, Joan L. **New York Times Book of House Plants.** New York: A & W Visual Library, 1975.

Free, Montague. **All About House Plants.** New York: Doubleday, 1946.

Jenkins, Dorothy H. **The Encyclopedia of House Plants.** New York: Bantam, 1974.

Schuler, Stanley. **The Winter Garden.** New York: Macmillian, 1972.

Van Alphen, Corry. **Effective Use of House Plants.** Buchanan, NY: Emerson Books, 1959.

Wright, Michael and Brown, Dennis, eds. **The Complete Indoor Gardener.** New York: Random House, 1975.

See page 117 for special sources on the topic of **Cold Weather Cooking and Eating.**

For further reading suggestions about automobile maintenance and repair, see page 144.

Illustration and
Photo Credits

Index